Introduction to Graphic Communication

by Harvey Levenson
and John Parsons

INTUIDEAS | RICOH

Published by IntuIdeas LLC in partnership with RICOH USA Inc.

Copyright ©2018 by Harvey Levenson and John Parsons. All rights reserved.

U.S. Library of Congress Control Number: 2018936401
International Standard Book Number (ISBN): 978-0-692-08117-4

Reproduction in any form, by any means, without express written permission is prohibited. Individual trademarks, trade names, and attributed product information are the property of their respective owners, supplied for information purposes only, and do not imply endorsement by or liability of the authors, IntuIdeas LLC, or RICOH USA.

This digitally-enhanced edition was printed in the United States, on a Ricoh Pro VC60000 inkjet press, by Edwards Brothers Malloy, Ann Arbor, Michigan, using 80# CVG Silk book stock and 118# 10 pt C1S cover stock with a gloss laminate finish, and perfect bound. Learn more about all this in Chapters 7, 8, and 9.

This book is dedicated to graphic communication professionals of all ages—those with experience and those about to embark on their journey into this ever-changing realm.

You are holding the first textbook to use Ricoh's Clickable Paper technology to access related digital media content and reader-to-reader interaction. Take a moment to download the free "CP Clicker" app on your iOS or Android tablet or smartphone.

To find the app store for CP Clicker, type this URL in your mobile browser:

igcbook.com/cp

As you read the book, scan the opening spread of each chapter with your tablet or smartphone, using the CP Clicker app. (There is also a URL for desktop use.)

Please explore the many learning and interactivity options we've added to this remarkable book. (The videos are on a secure, interactive learning portal—not YouTube!)

We will be adding new videos and other interactions over time. We hope you enjoy this new, hybrid approach to publishing, and we welcome your feedback.

book@intuideas.com
www.igcbook.com

INTUIDEAS | RICOH
Published by IntuIdeas LLC in partnership with RICOH USA Inc.

Table of Contents

Preface to the New Edition — 7

Introduction • About This Book — 9

A brief account of how this particular book was made, how to use its interactive features, and how it could change the way some books are published.

Chapter 1 • What is Graphic Communication? — 17

The start of a discussion—hopefully many discussions—on changes affecting the art, science, and business of printed graphic communication.

Chapter 2 • The History of Printing — 27

An overview of the 6,000-year history of innovation and disruption of printed graphic communication, and its importance as a foundation for our civilization.

Chapter 3 • Technological Transitions — 49

A brief look at the changing graphic communication business, including the impact of disruptive technology and innovation.

Chapter 4 • Print Industry Segments — 63

An overview of the different types of companies engaged in printing, how they overlap, and a summary of their current economic outlook.

Chapter 5 • Design and Prepress Workflow — 79

The practical aspects of digital design, preparation, prepress, and automated workflow steps that precede printing or publishing.

Chapter 6 • Color Management and Proofing — 97

The theory and practical aspects of managing printed color, achieving color consistency, and the proofing process—for both hard copy and virtual proofs.

Chapter 7 • Paper, Ink, and Toner — 113

The physical media of graphic communication: conventional and synthetic substrates, their sustainability, and the colored substances used to image paper.

Chapter 8 • Printing Processes — 131

A thorough review of the many ways ink or toner can be applied to a substrate, including both conventional and digital printing methods.

Chapter 9 • Postpress and Finishing — 159

An overview of the manufacturing techniques for folding, cutting, assembling, binding, and otherwise shaping printed pieces for consumer use.

Chapter 10 • The World of Packaging — 175

The many ways in which printed packaging differs from all other forms of graphic communication, and why this segment's growth will outpace all others.

Chapter 11 • Best Practices / Industry Standards — 191

A review of business practices and specifications for printing companies, as well as standards for increasing their overall efficiency and profitability.

Chapter 12 • Printing and the Digital World — 207

An overview of the Internet, the Web, and how the digital world has affected the business of printing—and our approach to graphic communication.

Epilogue, Bibliography, and Index — 228

Because a book with an online component is never "done," the Epilogue will be the starting point for future online discussions and reader-generated content.

About the Authors

Dr. Harvey Levenson is Professor Emeritus and former Department Head of the Graphic Communication Department at Cal Poly, and past Director of the Graphic Communication Institute. He remains active in the printing industry as a speaker, writer, consultant, and as an expert witness.

John Parsons is the principal of Seattle-based consulting firm Intuldeas. He is an independant writer, editor, and business analyst in the printing and publishing sectors. Until 2009, he was Editorial Director of *The Seybold Report*.

Preface to the New Edition

As with the original edition, this book is a survey, and a foundational overview of the complex world of graphic communication. It is focused on traditional and digital print, and the processes that make that possible. It is also about how digital technology, from desktop design to Web-based enabling tools, affects the art, science, and business of printing and imaging.

It is a snapshot of printing's past, present, and prospects for the future. It addresses the issues and challenges of printing today, to give the reader a better understanding of this often-confusing industry.

Introduction to Graphic Communication is written for those new to the field, including students in two- and four-year college programs in printing, print management, and visual design. It is also for high school and vocational students in comparable graphic arts and printing programs.

The book is also for internal corporate training and orientation—for professionals new to the industry or just unfamiliar with its peculiar nature.*

For those entering the field in the areas of sales, marketing, production, technology development, graphic design, advertising, and print buying, this book is a concise reference. For printing customers and service providers working with design and printing firms, it will serve as a guide, and help them make informed business decisions.

Whatever your field or discipline, graphic communication is everyone's business. Far from being dead or dying, print impacts the lives of nearly everyone in almost every human endeavor. It is changing, often with extreme consequences, but the fundamentals outlined in this book still apply.

This ambitious work would have been impossible without the generosity of others. Their contributions of talent, patience, and financial support have made this labor of love possible. At the risk of inadvertently omitting names, the following are individuals and companies to whom we owe an enormous debt of gratitude.

---

* Programs, companies, and individuals that do not require the full range of material provided in this book may purchase separately individual chapters on selected topics.

The Ricoh Commercial & Industrial Printing Business Group (Chris Reid, Eric Staples, Mike Herold, Annette McCary, Rob Malkin, David Bell, Akiko Tamura, and others) generously provided their technology and support, from the printing of the book itself to the Clickable Paper "bridge" to our learning portal. Steve Schulz and the team at Edwards Brothers Malloy also provided invaluable help.

Thanks go out to Tom Stine, Donna DeMarco, Carrie Strohl, Darryl Rentz, Bernie Selvey, Miguel Rodriguez, and the development team at Viddler for providing and supporting the online video portal for our interactive media.

Thanks also to Dina Vees of Cal Poly's Graphic Communication Department for her inspired book design.

Additional content, editorial feedback, and sage advice was provided by Brian Lawler, Frank Romano, David Zwang, Tim Ransom, Don Hutcheson, Timothy Baechle at Idealliance, and David Blatner.

Several prominent industry companies also provided generous financial support and content for the project. These include Sappi North America (thanks to Daniel Dejan and Patti Groh) and Silicon Publishing (Max Dunn). Thanks also to CGS Publishing Technologies International (Trevor Haworth, Marcus Goerlitz, Cory Sawatzki, David Palmieri, Bruce Brown, and Heath Luetkens), Markzware (Mary Gay Marchese and David Dilling), and Enfocus (Wim Fransen, Diana Albiol, and Toon Van Rossum).

Our Kickstarter contributors include John Renner, Heather Banis at Heidelberg, Kathrine and Nelson Mead, Barbara Ann Birkett, Bilge Altay, Daniel Dejan, Laura Roberts, Lois Lemon, Michael Stinnett, Robert Schaffel, Stephen Parsons, Tracy Campbell, Lucy Werner, Tim Vandehey, M. Licata, Thaddeus Kubis, Thomas Poon, Thomas Schildgen, Tom Stine, Twyla Cummings, Robert Lefcourt, Tom Carrig, Han Marshall, Thad McIlroy, Donna DeMarco, Laney Fugett, Marianne Nebel, Mark McCutcheon, Paul Bobnak, Paul Foster, Sarah Voigt, Tom Jory, Joanna Izdebska-Podsiadły, Don Carli, Lee Weir, and Ron Goldberg. Thank you!

Finally, special thanks go to Elin Bryce, Sam Parsons, and Warren Parsons for their welcome expertise in research, video editing, and user experience—and for their patience, perseverance, and support.

— *Harvey Levenson & John Parsons*

Introduction

Introduction

Preview

A New Kind of Book

Paper or Silicon?

How to Use This Book

Searching for a New Augmented Reality

Interactive Media: www.**igcvideo/Intro**

Web Links: www.**igcbook/Intro**

A New Kind of Book

Ten years after the first publication of *Introduction to Graphic Communication*, Harvey Levenson and I had a remarkable discussion. The original book, published by the Printing Industries of America, had sold well. Many schools and professionals were using it, but much had changed since 2007. An update was clearly needed, so we agreed to collaborate.

We wanted to blend online video and other media with the book—to make it a lasting, more engaging user experience. We thought: What better place to demonstrate the integration of print and digital media than in a book on graphic communication?

This would not be an eBook or an audio book, but an actual, physical book, printed on paper. It should also speak to the reader—to enhance the book's contents. The reader should be able to communicate with it, and with other readers. It should embrace multiple learning styles—including linear (reading), holistic (visual and sound), and collaborative.

The reader's experience with such a printed book had to be good, so we paid close attention to the narrative, illustrations, and overall design. Just as important, the online user experience had to match that of the book. It had to engage readers, not just play hours of video.

We made two important decisions. One was to host the videos in a secure, interactive learning portal. (More on that later.) The other was to connect the printed page to the online content, via a smartphone or tablet camera.

Rather than use QR Codes, we chose Ricoh's Clickable Paper technology. Both connect a printed page to a mobile experience, but they differ widely in scope.

Print provides a multi-sensory learning experience that digital media, for all its benefits, still lacks.

If a QR Code is a "doorway" to a single online destination, then Clickable Paper is more like Grand Central Station. It's a sophisticated, multi-faceted portal—to multiple, online destinations.

Instead of using a computer-generated symbol, Clickable Paper recognizes the actual page. Even a 500-year-old book can be the visual image that triggers an online experience.

This approach had never before been used with a textbook. But we believed that Clickable Paper's potential was enormous, particularly for books used in education and training. So, with Ricoh's support, and our first Kickstarter campaign, we started working on this ambitious, multimedia-enabled book.

Paper or Silicon?

Print is the "bedrock" or undergirding of this book. Conventional, zero-sum wisdom holds that the rise of digital media always leads to a corresponding fall in print. Proofs of a one-to-one tradeoff seem obvious, but the reality is far more complicated.

It turns out that people prefer a wide range of media types. Television changed the nature of radio, and how often we used it, but it did not eliminate it. Likewise, digital media have altered print, but will not eliminate it.

There is a reason why print is so persistent. It provides a multi-sensory experience. It is both visual and tactile. Touch, as it turns out, is an important aspect of how we experience and learn the world around us.

Print is not just a personal or generational preference. The book, *A Communicator's Guide to The Neuroscience of Touch*, by designer Lana Rigsby and neuroscientist David Eagleman, explored ways in which media shape the messages they carry. It highlighted decades of research in the field of haptics—the science of touch. Their findings are remarkable: **comprehension and long-term memory are significantly higher with print than they are with online media**. The study also found that paper *quality* can also affect how well a message is received, a response that holds true for all demographics—even "digital natives."

In 2016, renown designer Rick Valicenti also contrasted the properties of print and online media. He highlighted the multi-sensory qualities of printed paper, including not only touch but also its audible and olfactory qualities. Significantly, he also predicted a fusion of print and digital that could fundamentally alter how we communicate. We hope that this book (and the Rigsby/Eagleman findings) will prove him right.

When it comes to information discovery and immediacy, digital media have undeniable advantages. However, we believe that print must maintain a central role. This book explores the notion that the multi-sensory efficacy of ink on paper can effectively be combined with the versatility and flexibility of online media.

How to Use This Book

This volume is not an eBook—nor is it ever likely to become one. Instead, we created a parallel online experience, largely made up of curated video. Then, we connected each chapter of the printed book to interactive media, via the Clickable Paper app, so readers could access the content at any time on their mobile devices.

To create a better user experience, the videos are hosted on a virtual learning platform—the Viddler Training Suite—not on YouTube. This minimized distractions, added interactivity not found in other video environments, and kept the video content secure.

As you read this book, scan the opening spread of each chapter with your tablet or smartphone, using the free CP Clicker app.

To find the app store for CP Clicker, type this URL in your mobile browser:
igcbook.com.cp

Viddler was a good fit for higher education and industry training. Other kinds of books might require a different video platform. As with choices made on the print side, the experience should always fit the nature of the book.

Accessing the interactive video content is simple. Download the app to your mobile device, open it, aim your mobile device at the scanable page, and click on the finger icon that appears on your device. A choice of online experiences will appear. Unlike the QR Code or printed watermark approach, you do not have to aim your mobile device at a specific target on the page, only at the page itself.

For those who prefer a by-the-numbers approach, here are the details:

> ❶ **Download and install the free app.** Go to www.igcbook.com/cp for easy access to the iOS and Android versions. (This only has to be done once per device.)
>
> ❷ **Make sure you have a network connection** (either Wi-Fi or a mobile data network).
>
> ❸ **Use the app.** Go to the first two pages of any chapter. (Look for the "scan" icon in the corner.) Open the app. Using the camera feature, aim at either page of the spread, then tap the screen.
>
> ❹ **Find and interact with the related video.** In the Viddler portal, the videos are organized by chapter, and clearly identified with specific book content.

Not every page in the book is scanable, unlike the two-chapter prototype edition. (When the prototype was tested, users found that scanning every page was too burdensome.)

Readers also wanted to access interactive content on a desktop or laptop PC, so we added a short URL for each chapter.

In the Viddler portal, there are simply too many learning features to list—and more being added. So, naturally, there is a video that will help readers take better advantage of the environment.

Searching for a New Augmented Reality

We struggled with what to call this hybrid of print and digital media. Practically speaking, it's a book that "talks to you," via interactive multimedia, so we began calling it a "multi-book." It's also a type of augmented reality—without the fancy headsets and expensive 3D wizardry.

That said, what we call this new type of book is not important. What really matters is how it enhances the reader's journey, by combining the benefits of both media.

The possibilities for "print-to-web" or augmented books seem endless. For this volume, there are four types of interaction—with others in development. (The blessing, or perhaps the curse of digital media is that new things can always be added.) Here are the basics:

Interactive Video is the most prominent digital medium connected to the book. It is a series of "guest lecturers," demos, and other material, available on demand.

Group Discussions are available to all readers, initially via LinkedIn Groups.

Digital media are undeniably powerful tools for discovering and presenting information to a worldwide audience. They are also inherently chaotic. The process of collecting and curating video for this book proved how challenging it is to just "do a search."

Print is inherently stable and, according to research, better able to provide an effective, memorable experience. It is also long-lasting—on a shelf or some other, easily retrievable place. A book does not require additional electricity or tech support to be effective on its own.

Live Chat is also available, via an open Slack group. Private groups will follow.

External Websites including important online references, calculators, and demos will be added on a continuing basis.

Future online elements may include audio podcasts, reader blogs, and LMS interaction at colleges and even graphic arts companies using the book. (At the end of the book, we have added an Epilogue for the purpose of collecting and curating additional, reader-generated content.)

Rapid changes in mobile technology will undoubtedly make our approach seem quaint in the not-too-distant future. But we firmly believe that the fusion of these media types is inevitable.

The two should be combined. This book is an attempt to make a printed textbook the cornerstone—a reliable context for the volatile but valuable online world.

Printed books have been part of the human experience for over a millennium. From their block-printed origins in China, to Gutenberg's mass-production breakthrough, to the latest wrinkle in print-on-demand, books are a proven medium.

This experiment is based on the notion that digital media and interactivity is not a replacement for printed books, but a potential extension of a centuries-long history of innovation.

John Parsons & Harvey Levenson

Graphic communication answers are only a scan away...

Resources available by scanning this book include Web-based calculators, tools, and data tables for graphic communication professionals.

If your favorite calculator for color values, paper specs, binding styles, or other data are not there yet, please let us know: **book@intuideas.com**

1

What Is Graphic Communication?

What Is Graphic Communication?

Chapter Preview

The Complexity of Print

Digital Tools and Changing Roles

The Implications of Multi-Channel Publishing

Management Trends

A Shifting Model for Print Distribution

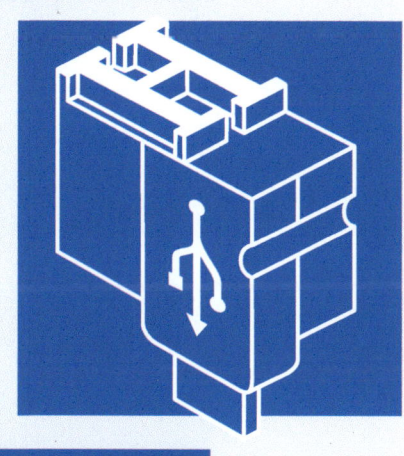

Overview

In this book, the term graphic communication refers primarily to the art, science, and business of print. As you will see, the practice of applying color to a substrate is more complex than it used to be.

Once considered simply a trade or a craft, printing today is a rapidly-changing business, connected not only to other forms of visual expression but also to every conceivable aspect of modern business. The emphasis on repetitive, manufacturing tasks has shifted to a more complex, technology-driven model, involving print as only one of many output channels. Today, graphic communication requires collaboration, strategic management, new creative and technical skills, and a re-thinking of roles and job descriptions.

This book focuses on print for very good reasons. One is that print, far from being in decline, is a growing part of visual communications as a whole. Despite this fact, however,

Introduction to Graphic Communication | 19

far too few graphic designers and professional communicators understand the basics of print.

This has had adverse consequences for the design community as a whole. In his 2013 journal article, *Graphic Design Curriculum and Printing Technology*, Dr. Bassam Al-Radaideh concluded that the two are inseparable. In his findings, the inclusion of printing technology education in graphic design curriculum contributes to the development of graphic designers, as well as offering them new capabilities and job opportunities.

Even in a mobile-focused economy, the absence of understanding of basic print reproduction is an unnecessary hindrance. Lessons learned in the print realm over many centuries—and focused in recent decades by disruptive technology—are extremely valuable to publishing, marketing, information management, and general business professionals.

Traditionally, printed graphic communication was divided into three categories: prepress, press, and postpress. Their core principles are comparable to those of Gutenberg's day. However, within a broader definition of graphic communication, we must explore new categories, and show how the traditional categories have been transformed by new technologies and practices.

This chapter will serve as an "overlay" for discussing how the industry has changed, and what new skill sets the industry will require. Each of the four sections outlined here can serve as a starting point for discussion—in class or online—of how our industry is changing and will continue to change.

When learning about specific aspects of print later in the book, it will be helpful to return to this chapter and ask again how these four trends have altered our understanding of graphic communication.

New Facets of a Traditional Process

At its core, graphic communication is about the mass dissemination of human knowledge—in the most reliable, durable form possible. That knowledge, however flawed or incomplete, is transmitted via commonly-used symbols and images, reproduced in forms that human senses can perceive and interpret, and on media that they can afford.

Print fulfills this role—in many ways better than broadcast or online media, as we will discuss in Chapter 12. However, it is undeniable that print itself has been transformed by the very technology that some claim (erroneously) will replace it.

SECTION 1
Digital Tools and Changing Roles

As with most fields, printing has been transformed radically by computer technology. Beginning with design and prepress (Chapter 5), there is no aspect of print production or business management unaffected by digital tools and infrastructure.

With new tools comes a radical change in job responsibilities, roles, and job descriptions. Tasks that were impossible to workers without computers suddenly became reasonable, and even commonplace. Innate skill was still necessary, but routine work was increasingly eroded by computerized automation.

To some, this trend represented a threat to well-established niches. Professional typesetting, for example, was subsumed by desktop publishing. However, in many cases, the rise of digital tools also created new specializations.

A prime example of this is the Electronic Publishing and Imaging Specialist. Typically, this person has a degree in graphic communication, graphic arts, or in a related field involving computer science. Experience in traditional print processes is extremely valuable, as is the ability to keep up with rapidly changing technology. (Graphic communication software and

DISCUSSION: Digital Tools and Changing Roles in Printing

In class or online, discuss the following questions:

- *What are the essential digital tools of an Electronic Publishing and Imaging Specialist?*
- *What are the traditional, print-related skills needed to be successful in this role? How are these skills combined?*
- *What are this person's responsibilities outside printing?*
- *How should this person's work be assigned value within a company? How should compensation be determined?*

Introduction to Graphic Communication

hardware undergo major upgrades every two to three years.)

Such a specialist is versed in the growing array of equipment and peripherals used in creating print and non-print media. This specialist is also adept in the use of software applications and how to optimize their use. Some may be involved training and leading others, and even in technology development and testing, either indirectly, as early adopters, or in positions within software or hardware companies.

Among other responsibilities, these specialists understand digital image preparation and manipulation, including retouching, color management, proofing, and color separation.

Additional their expertise may include a wide range of skills formerly performed by individual, pre-digital specialties, such as typography and page layout. Web authoring and interactive multimedia may also fall within this person's scope, as could image and document management.

Obviously, very few "Renaissance" individuals can encompass *all* the roles made possible by digital technology. However, the combining of skills and the blurring of former job categories is a fact of life in the digital age. While it has rendered some jobs obsolete, it has also created new opportunities for those adept at graphic communication *and* digital tools.

The imaging specialist is only one example of this trend. Other areas of print design, production, and fulfillment have seen the rise of similar, multi-faceted specialists. Very often, these individuals are not employed directly by a printing company, but by a publisher, advertising agency, design firm, or marketing entity.

In dealing with both print and digital media, digital software tools have created a new kind of workforce.

SECTION 2

The Implications of Multi-Channel Publishing

Design reproduction technology today has far greater implications than simply preparing artwork for print. Both images and text can be repurposed for multiple channels—each with its own technical and output requirements.

For example, photographic images for print use ultimately must be rendered as CMYK or process color, even though they originate in RGB, as discussed in Chapters 5 and 6. However, for online or digital publication, images are always expressed as RGB or hexadecimal data. Also, an image's required resolution (the amount of image data per square inch or centimeter) is significantly different for print than for on-screen use.

If print and online publishing channels were mutually exclusive, then choices would be a lot simpler than they are. Each medium would have its own set of requirements. Professional skill sets and technologies would evolve independently. That is not the case. Publishers and other communicators nearly always use multiple channels to tell their story. An image must be useable in more than one environment—a phenomenon known as repurposing. This has led to the rise of a new professional role: the Design Reproduction Technologist.

DISCUSSION: The Implications of Multi-Channel Publishing

In class or online, discuss the following questions:

- What are the essential software tools—and corresponding skill sets—of a Design Reproduction Specialist?
- How would you most efficiently repurpose a print project to an online version—or vice versa?
- What are the characteristics of an "output-neutral" data repository? What is needed to create print and online content from such a source?
- What are the team requirements of content repurposing?

Individual designers may be aware of these issues, but usually cannot manage all the implications of image repurposing. It is the role of Design Reproduction specialists to focus on technology solutions, freeing designers to focus on the creative aspects of their work.

Multi-channel repurposing is not limited to images. Text for print applications today is governed by typographic styles and formatting choices contained in both word processing and desktop publishing software—as discussed in Chapter 5. For designers, these choices are as important as the text itself.

However, in the digital world, formatting and even font selection is an entirely different process. It is often the Design Reproduction Technologist's role to provide a transparent (to the designer) means of visually mapping the typographic design intent from one environment to another. In addition, text and images may reside in a central, "device-neutral" data repository. It then becomes the job of a specialist to place that content in multiple channels—both print and digital—in templates created by visual designers.

Effective repurposing is not a manual process, even by a highly-skilled Design Reproduction Technologist. They must deploy many of the systems and processes described in Chapter 12.

Although the technical requirements of such a specialization are rigorous, they must also be cognizant of design principles and expectations. Depending on the size of the organization, such a specialist may be involved directly with the client, from developing original designs, anticipating online or mobile interactions, and carrying out production tasks to make the idea succeed in multiple venues.

DISCUSSION: Changes in Printing and Imaging Management

In class or online, discuss the following questions:

- How has the shift from a manufacturing-centric business to a full service model affected management style?
- Can printers succeed in offering non-print services to clients with multi-channel communication needs?
- To what extent should management become collaborative or team-based in a more complex print environment?

SECTION 3
Printing and Imaging Management Trends

Perhaps nowhere have job requirements changed more radically in today's multi-channel world than in the management of graphic communication companies. Once focused on the production of a single class of products, printers have had to diversify. This includes not only an expansion of the types of printed product they create, but also an array of digital output and ancillary services.

Managing such a business is challenging. A formerly traditional printing company must compete with digital-only producers, agencies, and marketing firms. It must also deal with increased competition for basic printing services, thanks to online ordering and a tendency to view print as a price-based commodity.

Effective printing and imaging management involves a knowledge of multiple technical aspects of print and non-print media, as well as the traditional requirements of strategic planning, financial acumen, marketing, sales strategy, supply chain logistics, and team leadership. Obviously, these qualities can never be found in one person. So, in all cases, finding key players—and ways to work collaboratively—is paramount.

SECTION 4
A Shifting Model for Print Distribution

For centuries, the **print-and-distribute** model was the only possible option for putting printed pieces in the hands of consumers. Printing equipment was specialized and costly, requiring it to be located centrally—typically near urban centers or where labor and sources of energy were plentiful.

Once they were printed, books and other material required an elaborate distribution network to supply retailers and end-users. To reach an individual reader, a printed piece had to be shipped, stored, and handled multiple times.

Improvements in supply chain efficiency made the print-and-distribute model more tolerable. The cost of distribution also drove the printing industry to innovate. Lighter-weight, high-quality substrates and distributed warehousing are two pre-digital examples.

However, the real change in distribution occurred in the wake of radically new telecommunications technology. In the 1960s, Dow Jones began transmitting the entire *Wall Street Journal* electronically, via satellite, from New York to remote locations for printing and regional distribution. It was an early example of the **distribute-and-print** approach.

This was not a new concept. 19th Century telegraphy and the "telefax" had already made text and even image transmission a reality. However, the Dow Jones milestone proved that entire publications could be electronically distributed and printed much nearer to the end user.

The distribute-and-print model is ubiquitous today. It is second nature to send someone a PDF or other digital file, where it is often printed out, bound in some fashion, and then delivered.

However, the same model applies to information that requires more sophisticated printing and binding. Print-ready files are sent to local or regional printing facilities to shorten delivery timelines, to reduce inventory and shipping costs, and to decrease the environmental footprint of shipping from a single location.

This shift has been made more complex by the introduction of non-print media. However, The Internet has not reduced total print volume. Rather, it has fostered a trend towards more efficient, decentralized production. Even in packaging, a medium not threatened by the Internet, printing process occurs increasingly at the product manufacturing location.

This fundamental change in distribution models will continue to make graphic communication more complex—requiring new skill sets and specializations. Far from being a boring or obscure aspect of the field, print distribution is essential to a firm understanding of the changing world of graphic communication.

DISCUSSION: Implications of the Shift to Distribute-and-Print

In class or online, discuss the following questions:

- What are the business advantages of distribute-and-print?
- How has the PDF standard affected print distrubution?
- How will Net Neutrality and other telecommunication policies affect print distribution?
- How has the Internet changed futurists' predictions of printing our own, customized publications at home?

2
The History of Printing

The History of Printing

Chapter Preview

Printing & Human History

The Substrate Revolution

Before Gutenberg

The Renaissance

Printing in America

19th Century Print Industrialization

The Rise of Lithography

The Electronic Era

Overview

Human beings have embraced graphic communication since prehistoric times, using visual elements to express ourselves and tell stories. Before about 4000 BC, we did so only one piece at a time. Whether a carving, a painting, or marks on a piece of clay or sheepskin, each piece unique. It could be shared only with one person or group at a time. The beginnings of civilization coincide with our discovering ways to create copies of our graphic communication more easily, and share those visual elements with a larger audience.

Throughout this book, we will focus on the cornerstone of modern graphic communication: print. But printing is not just ink or toner on a sheet of paper. In its broadest definition, printing is the reproduction or duplication of visual elements on a surface—any surface. From carved stamps on 6,000-year-old clay objects to colored pixels on a touchscreen, we all "print" text, images, and other media on a substrate—in order to tell our story to others.

As a manufacturing industry, printing dates back to 1456, with Johann Gutenberg's famous invention. He did not invent printing, or even the printing press, as is popularly believed. Gutenberg's breakthrough was the process of duplicating movable type. For the first time, he made possible the mass production of print.

Before Gutenberg, books and other documents were laboriously produced by hand. By creating molds for cast metal type characters, and using them to print pages on a modified wine press, he automated the

Introduction to Graphic Communication | 29

process of designing and reproducing the printed page in large quantities. This accelerated the production and dissemination of the printed word. Today, the printing industry still has the same goal: to mass-produce and distribute documents as quickly and efficiently as possible.

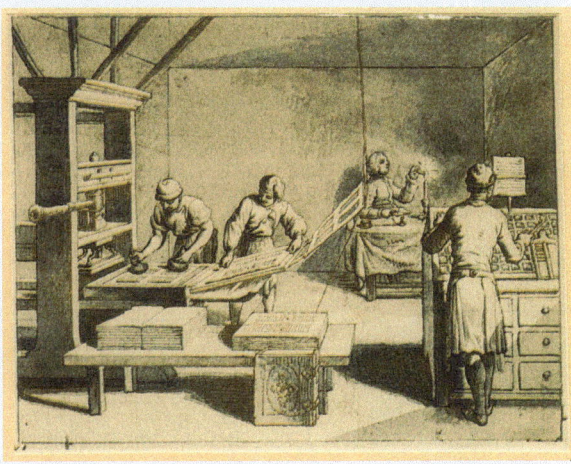

Gutenberg's invention is important for another reason. It typifies the increasing disruption we see today. Before Gutenberg, changes in graphic communication were infrequent. After Gutenberg, the "curve" began to get steeper. To be sure, his typesetting method was unchanged for 430 years—supplanted in 1886 with Ottmar Mergenthaler's invention of the Linotype machine. But by 1970, rapid change became the rule, not the exception.

This chapter will give you a better grasp of the significance of printing and its role in advancing human civilization. This is especially true of the last two millennia. The printing press and moveable type have been ranked as the world's most influential invention—topping gun powder and the compass. That influence has grown, with the accelerated pace of disruption, and the expansion of "printing" to all forms of physical and digital graphic communication.

Graphic Communication, Printing, and Human History

The story of civilization is the story of communicated ideas. Man has long been able to express ideas visually—on stone or animal skin—but only on a single artifact. Communication with a larger audience is essential for every civilized institution. This requires that our graphic expressions be replicated, which is the story of printing. Institutions like education, law, religion, or medicine would simply not exist in their present state had it not been for the dissemination made possible by printing.

The first reproduction of graphic images is believed to have happened between 4500 B.C. and 3500 B.C., with the use of carved stone or metal "stamp" seals.

Prehistoric cave painting in Lascaux, France

These were used to indent ownership marks into moist clay. (This could be considered the first attempt at visual branding or, literally, trademarking.) About 3500 B.C., cylinder seals containing duplicate relief, or raised, symbols.

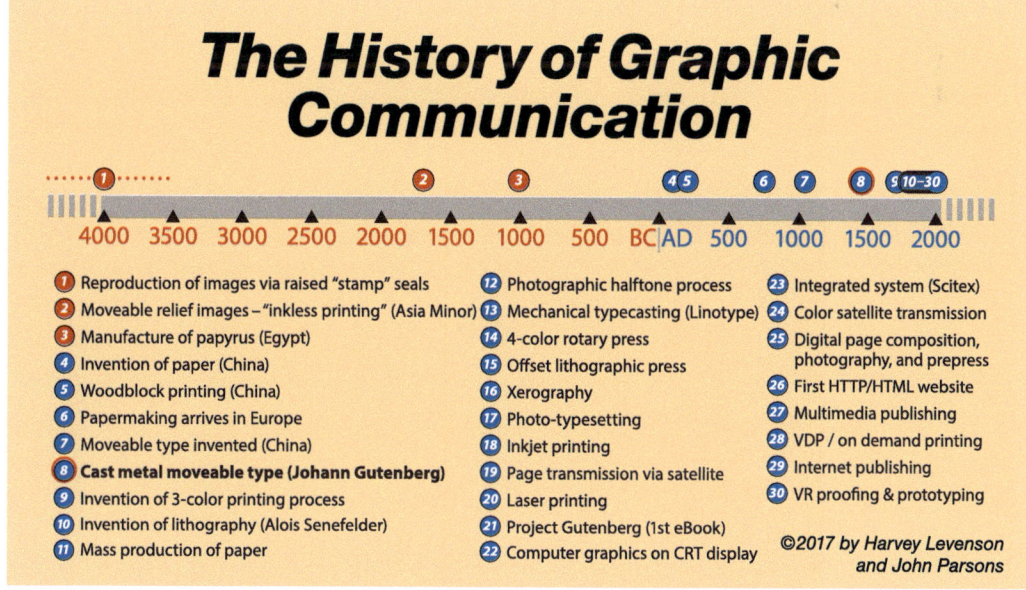

Graphic communication—the reproduction of images and text on a practical medium—spans six millennia of human history, from ownership marks stamped in clay to mass production of visual elements on physical or digital surfaces.

Introduction to Graphic Communication | 31

The Phaistos Disc
(Heraklion Archaeological Museum, Crete)

Other early image reproduction was largely done using clay as a medium. Movable relief images for "inkless printing" are believed to have been used in Asia Minor around 1700 B.C. Type-like relief symbols were impressed into soft clay, such as the circular pattern of syllables found on the "Phaestos disc" on Crete. This suggested an alphabetic structure, and the first known use of re-usable images, comparable to Gutenberg's relief type over 3,000 years later.

The Substrate Revolution

Imprinting images on clay has obvious limitations. What eventually led to printing as we know it is the development of a cheaper, lighter, more versatile medium (or substrate) on which to place our text and images. The first such substrate was papyrus.

The history of communicating with the written word began between 1085 B.C. and 950 B.C. with the manufacture of papyrus and the use of pictography, or "picture words." While the ancient Egyptians had many uses for papyrus, its real importance was in the manufacture of what was then considered "paper." For centuries, it was the primary means of conveying pictographs and written words. These were created manually rather than by printing.

Papyrus reeds grew along the banks of the Nile River. In making paper from papyrus, the fibrous cores of stems were first cut into strips. The strips were then pounded, laid crosswise into a single sheet, pressed together, and dried. The sheet was then burnished with a stone to create a smooth surface, and a reed pen was used to write or draw on its surface.

Although vastly superior to clay, and cheaper than animal skin, papyrus-based paper was not the ideal substrate for writing or, eventually, printing. The real revolution came from China in A.D. 105, when Han dynasty court official Ts'ai Lun discovered how to make paper from a pulp mixture of mulberry bark, hemp, linen rags, and water. The mixture was spread on a porous screen or mat, the water pressed out, and the resulting sheets dried in the sun.

Ts'ai Lun

The Muslims carried the art of papermaking to Europe in 751. However, it did not change fundamentally from Ts'ai Lun's original wood fiber process for almost 19 centuries. The increased demand for paper, beginning with the growth of printing beginning with Gutenberg's invention, and more recent moves to overcome raw material shortages, have fueled the need to innovate and discover substitute materials for papermaking, creating substrates with the strength, longevity, and surface characteristics needed for good graphic reproduction.

Ts'ai Lun's "pulping" process for making paper is still used today—on highly automated machines costing tens of millions of dollars. While paper often is still made from wood pulp, cotton, flax, and other seed fibers are also used. In addition to the fibers themselves, paper includes an increasingly complex array of coating chemicals, dyes, optical brighteners, and strengthening agents.

The manufacture of paper was improved with the invention of the Hollander beater—a device for beating rags to a pulp—around 1700. In 1797, French paper mill clerk Louis Robert designed a paper-manufacturing machine that was commercially developed by English engineer Bryan Donkin from 1803–1812. His project was financed by prominent paper manufacturers Henry and Sealy Fourdrinier, so the machine has been called the Fourdrinier ever since. While it has been improved over the years, the Fourdrinier still remains essentially the same, and is the dominant papermaking machine in use today. Modern Fourdrinier machines can be several hundred yards long, and can make paper at speeds up to 3,000 feet per minute.

Today, the development of synthetic paper, with no pulp required, is becoming mainstream, Indeed, the importance of paper seems to get stronger with every suggestion of a "paperless" society. The promise of the 21st Century is not of paperless communication, but of a process that embraces the quality of paper in new and creative ways.

Mixing pulp has always been part of the paper making process.

The term substrate should not be limited to paper. Printing can be done on almost any material. In addition, the development of "digital paper," by MIT Media Lab and Xerox Palo Alto Research Center (PARC), is as significant as Ts'ai Lun's discovery. As discussed later in the book, a digital substrate can display both text and images, and be electronically re-imaged with new text, once the original is no longer needed.

From its origins in Egypt and China to its more recent mass production and diversification, the convenient material we call paper is the foundation of graphic communication. As a durable, versatile, and disposable commodity, paper is synonymous with the growth of printing.

The First Printing Revolution

Johann Gutenberg did not invent the printing press, nor did he invent moveable type. Although he revolutionized printing, the process predated him by over 1,000 years.

Hand-carved woodblock printing was also first used in Asia. Historians credit the Chinese with woodblock printing on paper at least as early as A.D. 400. However, the Japanese Empress Shotoku commissioned the first "mass market" publication, a million copies of woodblock-printed sheets, between 764 and 777. Called the "Million Buddhist Charms," the collection of prayers is thought to be the oldest surviving example of woodblock printing in Japan, and the earliest known specimen of printing. In 868, the first block-printed book, The Diamond Sutra, was produced in China.

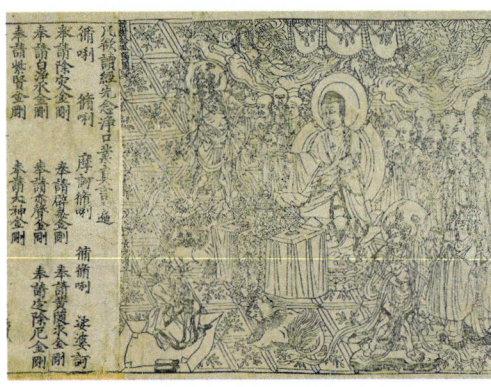

The Diamond Sutra

Movable type for printing was also invented in Asia. A Chinese alchemist, Pi Sheng, invented type made from baked clay about 1043. A font of 60,000 wooden types was produced for the Korean ruler Wang Chen in 1313. Success of these type fonts was limited because of the large number and complexity of pieces required for printing Asian languages. Hence, there is evidence of metal type being used in Asia by the 14th Century and of the establishment of a central department of books.

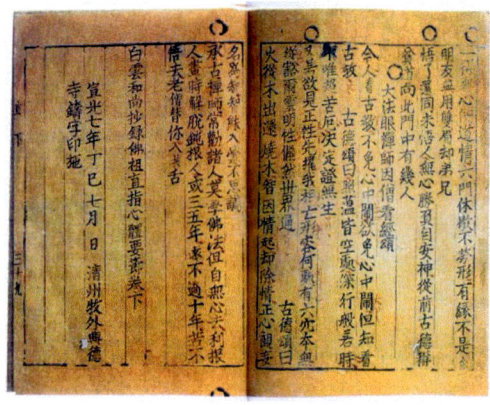

Jikji, a Korean book of Buddhist teachings, the earliest known book printed with movable metal type, in 1377.

It is widely believed that Buddhists introduced printing into Korea. The Korean king Ta-jong (1401–1419) was the first to carry out the idea of movable type made of copper. By 1403, forty-seven years before Gutenberg's first printing, Korean type was also being cast in bronze.

The first book printed in Japan with movable type occurred in 1596. However, because of the complexity of Asian alphabets, printing from woodblocks was much more practical and less expensive than printing from movable type.

Gutenberg's invention—one that forever changed printing from a manual craft to an industrial process—was cast metal moveable type. The art of typography is also attributed to Gutenberg, who designed his typefaces to simulate the hand-lettered text of his time. His invention survived for nearly 500 years before it was replaced by the Linotype machine (followed by phototypesetting and then by computerized type.)

Movable type worked perfectly for Gutenberg, for he had only the twenty-three basic letters of the Roman alphabet to deal with. He did more than just recycle the Chinese idea; he invented a mold that could turn out metal type characters of exactly the same height. Thus, when the type was assembled, inked, and pressed against paper, the result was clean, uniform, and highly readable. Gutenberg could produce a hundred copies of a page in the time it took a scribe to make one or two pages.

Other printers in Europe were also experimenting with printing processes, but Gutenberg was the only printer known to be using movable metal type. Theirs was a 15th Century race to develop the first workable method for producing and reproducing movable type as a logical extension of manuscript writing, and to serve a growing demand for information Europe. Gutenberg's method of creating type by pouring molten lead into precise molds seemed to work best. (His first work was actually not the Bible, but a mass-produced form: the infamous papal indulgences that helped spark the Protestant Reformation.)

Gutenberg's breakthrough was the casting of type characters in lead, using a mold created from a harder, engraved metal.

Between 1450 and 1455, Gutenberg and his assistant, Peter Schoeffer, printed approximately two hundred copies of what has come to be known as the "42-line Bible," of which only 49 copies still survive.

Gutenberg's press was designed from a 15th Century wine press. The press was not the essence of His invention was a necessary accessory.

After Gutenberg's invention, for the first time, hundreds of readers owned identical copies of the same book. A book one read in one city was the same someone else was reading in another city. It also generated the first case of "information overload," as books proliferated. Just

A page from the Gutenberg Bible
(History of the World)

Introduction to Graphic Communication | 35

as rapid social change parallels today's digital information explosion, so did chaotic changes like the Reformation follow Gutenberg's breakthrough.

Ts'ai Lun's and Gutenberg's inventions altered the structure of institutions and influenced how learning took place. They defined printing as the most meaningful, detailed, pervasive, and informative form of mass communication.

The disruptions following Gutenberg have continued to this day. In the 16th Century, French Humanist Peter Ramus introduced the concept of print-oriented classrooms in which books were available to all students. Before printing, much of the time in school was spent creating texts; the classroom tended to be a "scriptorium" with the student as editor-publisher. Printing changed learning. The book was the first "teaching machine" and also the first mass-produced commodity.

A woodcut of Gutenberg's press
(History of the World)

Understandably, the proliferation of print put scribes out of work. Their plight resembles that of linotype operators in the 1960s with the demise of linotype machines. It also foreshadows almost every other labor disruption we see in graphic communication today—from the Internet-enabled automation of the pressroom to the rise of online publishing and much more.

Gutenberg's disruptive innovation ended the centuries-old model of producing books in a "Scriptorium."

The printing industry has also experienced "concept-to-market" success delays with the introduction of nearly every new technology. The Linotype machine was introduced in 1885, but was not considered a commercial success until 1905. Electronic color separation scanners were available in the 1940s, but did not become commercially viable until the 1970s. It took the Gannett Corporation more than ten years to realize its first profits using satellite transmission to produce USA TODAY in color at multiple sites simultaneously. The concept and technology of on-demand digital printing has created enormous market anxiety for those who have invested in these technologies and are attempting to develop niches for them. (These technologies are covered elsewhere in this book.)

Over the centuries, the development of each new print technology has created new challenges, eventually forcing graphic communicators to adopt new models. For example, books flourished as a result of the development of copper engravings, but the transition was not an easy one. Copperplate engravings were considered superior to woodcuts by the late 16th Century, but they required a special, more expensive printing press. (Woodcuts could be inserted in the same print form as the type characters and printed on the same press.) However, with the rise of other printing processes, images could be engraved in the same copper plate as the type characters, and printed on the same press.

That particular process was first developed about the same time Gutenberg was perfecting his invention for relief or letterpress printing. Called "intaglio," this process employed the opposite principle of printing from an engraved or depressed surface, and was used for copper engraving as early as 1446. Metal smiths produced intricate designs by carefully inscribing or scratching depressions into a soft copper surface. Ink placed on the finished engraving and wiped off the top surface would remain in the recessed depressions; this ink could then be transferred to paper pressed against the image. (Today's steel engraving and rotogravure processes use the intaglio principle.)

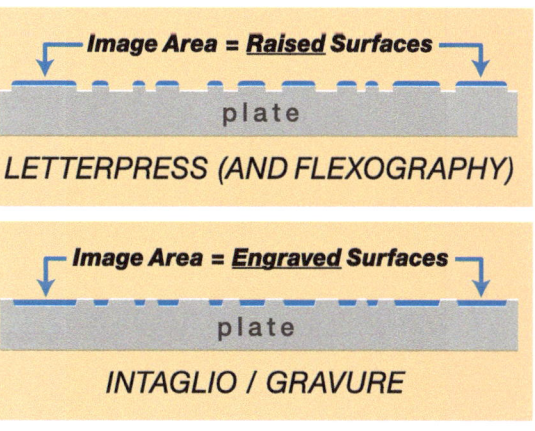

The two original printing processes used two opposite methods, applying ink to either raised or engraved surfaces.

As we'll discuss later in the chapter, the rise of alternatives to printing with raised and engraved surfaces are significant. However, the importance of the first printing revolution cannot be underestimated. As epitomized by Gutenberg himself, this shift altered the use of language as a means of perception and exploration, and made them a portable commodity. Printed books were the world's first uniform, repeatable, and mass-produced items, although the full production-to-consumption implications were not fully realized until the 19th Century. That fundamental shift towards commoditization pervades all graphic communication today—from modern printing on paper to every form of digital media.

Introduction to Graphic Communication | 37

The Importance of Type

With the advent of printing technology came the need to develop type designs that reproduced properly on a printing press within the limitation that press technology imposed. An image produced by squeezing ink on paper was different from one produced by writing on paper. Thus, from its origins in handwriting, altered by the needs of mass production, the art of typography and the science of typesetting emerged.

Typography relates to the aesthetics of type and type design while typesetting refers to the technical operation of composing, setting, or arranging type for printing. The latter has changed significantly over time, from handmade letters originating in Asia, to Gutenberg's breakthrough in cast metal, to photographic and digital processes today. Regardless of the technology, however, the use of typographic elements is central to all forms of graphic communication. In 1470 Nicholas Janson, a famous Italian printer and

Janson
ABCDEFGHIJKLMNOPQRSTUVWXYZ
abcdefghijklmnopqrstuvwxyz 1234567890

Caslon
ABCDEFGHIJKLMNOPQRSTUVWXYZ
abcdefghijklmnopqrstuvwxyz 1234567890

Baskerville
ABCDEFGHIJKLMNOPQRSTUVWXYZ
abcdefghijklmnopqrstuvwxyz 1234567890

Bodoni
ABCDEFGHIJKLMNOPQRSTUVWXYZ
abcdefghijklmnopqrstuvwxyz 1234567890

Goudy Oldstyle
ABCDEFGHIJKLMNOPQRSTUVWXYZ
abcdefghijklmnopqrstuvwxyz 1234567890

typographer, perfected the clarity, beauty, and utility of the Roman typeface—a monumental contribution to the history of the graphic arts. Janson's first Roman typeface introduced thick and thin strokes and serifs. His alphabets—based on simplicity and appealing stroke and space proportions—still influence type designed more than five hundred years later.

Thousands of type styles exist today, some evolving from the work of a few 18th Century designers. Three great names in early type design are William Caslon and John Baskerville of England and Giambattista Bodoni of Italy. Each developed widely acclaimed type designs that bear their last names and became the forerunners of major divisions of Roman type used for printing.

A more contemporary 20th Century typographer was Frederick W. Goudy. During his lifetime Goudy designed more than one hundred typefaces, considered by some experts to be the most beautiful in existence.

The rise of digital publishing has given new significance to the art and science of type. The basic elements of good type—readability and aesthetic appeal—remain the same. But reading on screen (using a transmitted light source) is different from reading on paper (using a reflected light source). Modern day Jansons, Caslons, and Goudys will have to create type that interacts well with the human eye in many new ways.

The Printer's mark of William Caxton, 1478

Printing During the Renaissance

During the Dark Ages, rulers and clerics kept knowledge mainly for themselves. Gradually, however, the general public became more involved in art, history, and science. This rebirth of learning became known as the Renaissance. Professional manuscript writing had become an industry during this era. But the manuscript industry was doomed by the rise pf printing, which further propelled the rebirth of learning throughout Europe and beyond. Printing became an important aid for dissemination of knowledge. It advanced persuasion over unquestioned belief and became an essential tool for informing the masses.

Printing quickly spread beyond Gutenberg's Mainz, in Germany. William Caxton brought printing to England in 1476, a year after he had printed the first book in English in the Netherlands. (Caxton is often blamed for today's difficult spellings of certain English words, like "laugh" and "rough." Had he printed a few years later—after the shift from Middle English to Modern English—he would have represented words the way they sound today.)

One of the greatest figures in publishing history, Aldus Manutius established a printing company in Venice in 1495. Although his works are distinguished by

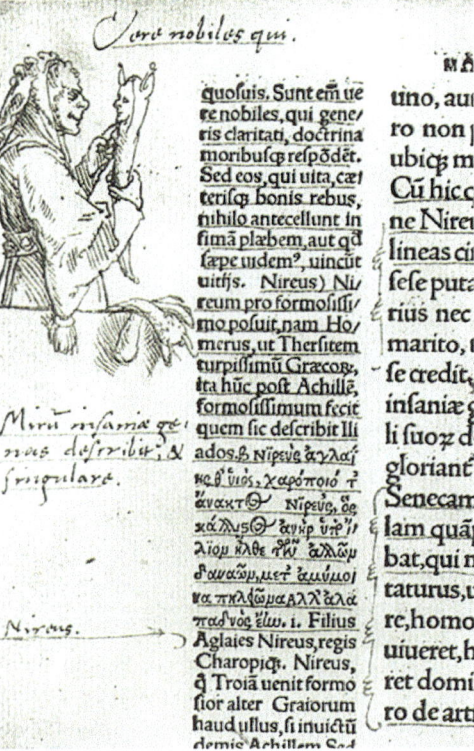

Marginal drawing of Folly by Hans Holbein in the first edition of Erasmus's Praise of Folly, 1515

exceptional typographical beauty, Manutius wished to make it possible for common people to afford editions of Greek and Latin classics. He published "pocket-sized" works of Aesop, Aristotle, Euripedes, Homer, Plato, Plutarch, and others at a low cost. In 1501, he printed the first book containing italic type.

Dutch theologian and philosopher Erasmus, responsible for the first printing of the New Testament in 1516, used the emerging print technology as a way of expounding ideas through printed words. One of his chief interests was the republication of works by Aldus Manutius.

While printing's initial influence was largely on the humanities, the technology of printing eventually shaped the sciences as well. Modern sciences depend on information conveyed by exactly repeatable visual or pictorial statements.

During and immediately after the Renaissance, the printing of books was almost entirely dependent on the raised surface printing technique (letterpress) used by Gutenberg. Single-sheet works—called broadsides—were also common. Newspapers were also produced via letterpress, beginning about the 17th Century.

Printing Comes to North America

The first printer in North America, Juan Pablos, arrived in 1539 and set up his press in Mexico City. Other printers followed and Mexico became a printing center in North America. But it would be a hundred years before printing reached what is now the United States.

The first known printing in the British Colonies was a one-page "Freeman's Oath," produced about 1638 by Stephen Daye in Cambridge, Massachusetts. The first book printed in Colonial America was a book of psalms printed by Matthew Day

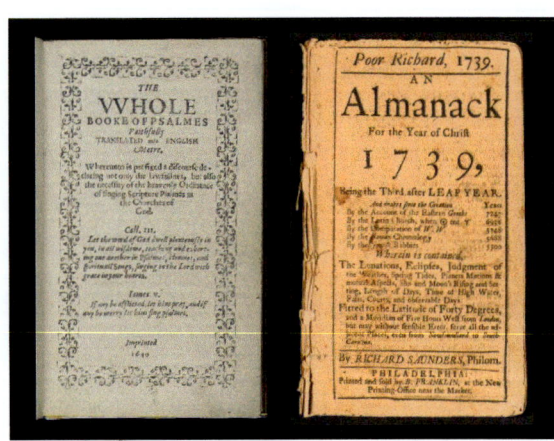

A New England book of Psalms, 1640, and Franklin's Poor Richard's Almanac, 1739

in 1640. Later came prayer books, almanacs, bibles, sermons, lesson books, and newspapers—all produced on wooden presses using Gutenberg's now-traditional process.

As printing grew in New England, so did the need for a ready supply of paper. William Rittenhouse started the first paper mill in 1690 near a small river outside Philadelphia.

As it had in Europe, printing in the Americas had a profound impact on its religious and political life. By the time of the American Revolution, literacy was at an all-time high, and British military officials complained that the Colonists' love of books had undoubtedly made their willingness to fight over principles even stronger.

Early Newspapers in America

The Boston News-Letter, published in 1704, is considered the first American newspaper, although Benjamin Harris had printed a single issue of a one-page Publick Occurrences in 1690. Established by John Campbell and printed in the shop of Bartholomew Green, the News-Letter was published until 1776.

Benjamin Franklin was a skilled scientist, author, editor, publisher, and printer. His Pennsylvania Gazette, started in Philadelphia in 1732, was a well-respected newspaper. Among Franklin's many writings was the series of Poor Richard's Almanacs, published annually between 1732 and 1758 in runs as large as 10,000 copies on a hand press.

In 1814, after centuries of using history's oldest printing machine, the hand press was finally superseded and virtually discontinued forever as a production tool when Friederich Koenig sold his first power-driven press to The London Times. Koenig's invention immediately lowered printing and composing costs by 25 percent and made possible the production of inexpensive and longer publications of every kind.

J. J. Lankes' 1845 Hoe Washington Handpress (Image courtesy of the Tampa Book Arts Studio at the University of Tampa.)

In 1822 Peter Smith, an American associated with the R. Hoe & Co., devised a machine that was in many ways superior to any concept up to that time. Its frame was of cast iron, and in place of a screw with a lever he substituted a toggle joint, which was simple and effective. In 1827 an invention by Samuel Rust was a great improvement on the Smith press. Instead of being all cast iron, the press frame's uprights at the side were hollowed for the admission of wrought-iron bars, which were securely riveted at the top and bottom of the casting. This gave additional strength and greatly diminished the amount of metal used in construction. The new invention was known as the "Washington Hand Press," and in principle and construction has

Hoe's six-cylinder rotary press

never been surpassed by any hand printing press. The Washington Press was manufactured in great numbers and sold around the world—only to be superseded by the universal adoption of the cylinder press. More than six thousand Washington Presses were made and sold.

Harper's Publications

From the mid-1800s to the mid-1900s, a series of concurrent developments greatly widened public access to print. Steam-powered cylindrical presses, stereotype plates, and paper mass-produced from wood pulp greatly increased the output and decreased the price of printed material. By the end of the 19th Century, several hundred times as many pages were printed per capita as at the beginning of the century. The invention of the telegraph and the laying of the Atlantic cable enormously enhanced the speed and efficiency with which the press could convey news. Additionally, the completion of a rail network made practical the swift nationwide distribution of magazines and books. Nearly universal elementary education and widespread literacy opened a broad public market for print. With the "penny newspaper" and inexpensive magazines and books, print became a mass medium.

Mass production of printed material became easier in the 19th Century through a series of improvements in press designs. The first power-driven cylinder press sold to The London Times in 1814 speeded printing by carrying paper in a circular fashion around a cylinder to contact a flat form of type below. In 1846 R. Hoe & Co. produced a type-revolving press—the first to use a rotary, cylinder-to-cylinder, principle. Its design required that forms containing loose pieces of type be wrapped around one cylinder. This design was replaced in 1871 with a rotary press that used single-piece stereotype plates. George P. Gordon devised a treadle-operated version of a platen press in 1850.

In 1856 William Bullock produced the first web-fed, perfecting press, capable of printing on both sides of a continuous roll of paper.

Charles Mahon, third earl of Stanhope, re-established stereotype printing in England and perfected his method of stereotyping from plaster molds. His process offered obvious advantages wherever books were being frequently reprinted such as bibles, prayer books, and schoolbooks. Stereotypes greatly reduced the wear of type, and a stereotype plate would not break into pieces as a form of type often did.

The first line of Gordon job presses was put on the American market by George Phineas Gordon in 1851 and was used for efficient printing production well into the twentieth century.

The Linotype Machine
(Image courtesy of Dr. Bernd Gross.)

The Linotype Machine

The Industrial Revolution contributed one of its most significant technologies when on July 3, 1886, Ottmar Megenthaler, a German immigrant, demonstrated to the New York Tribune a machine that today is regarded as one of the ten greatest inventions of all time: the linotype machine. Type for nearly every newspaper in the world and most books in the United States would eventually be composed on Mergenthaler's machine or on machines that achieve similar results such as the Harris Intertype or the Monotype machine.

When keys on a linotype are struck, pieces of brass punched with characters are brought together into a line, automatically spaced, and moved to a mold where molten metal is injected into the indented characters to produce slugs of metal type in complete lines. The Harris Intertype operated on the same principles, whereas the Monotype machine created type from individually cast characters automatically formed into lines. The Monotype had the added distinction of being driven by punched paper tape produced on a keyboard that was separate from the typesetter or caster. Punched paper tape was used in this capacity for decades before the same concept was applied to computers.

The linotype and similar technologies completely revolutionized printing and publishing, allowing far more pages to be printed in newspapers and magazines than was possible with handset movable type. Today, publishing still depends on "keyboarding," albeit through computers. The linotype machine was completely obsolete by the mid-1960s.

Photographic typesetting, typically controlled by a computer system, replaced hot metal type with "cold type," imaged on photographic paper. These were themselves replaced by desktop publishing systems on personal computers.

The first practical typesetter that did not require molten hot metal to produce images was the Fotosetter tested by the Harris Intertype Corporation in 1946. While this machine's outward appearance resembled the hot metal linecasters of the day, its imaging system was quite different. The small brass matrices (mats) that circulated inside the machine contained not tiny letter molds but film negatives, each with an image of a letter, numeral, or symbol. As the matrix passed a specified point in its travel through the machine, a beam of light flashed through the negative in its side, exposing the image onto photographic paper or film. The Fotosetter was soon displaced by more advanced and faster photo-optic systems of simpler design. Hence, a synergy of photography and typesetting became the mainstay of typesetting technology until the mid-1980s, when desktop publishing took over the typesetting function for most printing applications.

Before the 20th Century, letterpress—greatly mechanized and automated by inventions like the rotary press and the linotype—dominated a large percentage of the print industry. Gravure (intaglio printing) was second, but there were few alternatives.

The Rise of Lithography

In 1789, Prague-born Alois Senefelder invented a process that would alter the world of print: lithography. The term means "stone printing," and comes from the use of the smoothed surface of a porous limestone slab, upon which images for printing were drawn, engraved, or transferred. Although the printing aspect of lithography soon moves to other imaging surfaces, stone lithography still has a place as an art form for original printmaking.

Lithographic stones were made from porous Bavarian limestone upon which images were drawn, engraved, or transferred through a chemical process that separated the image and nonimage areas.

Alois Senefelder (1771-1834)

It is based on the principle that oil and water do not mix. Besides inventing special, greasy ink for the process—which was used to write or draw on the stone in reverse—Senefelder also devised a method for transferring the images to a special paper which he called a transfer sheet. Hence, he was able to draw or write a "right-reading" image that was then transferred—or offset—as a "wrong-reading" (or reverse) image on the stone. The image would then be transferred again as "right-reading" onto the paper.

To make prints from the image, Senefelder sponged the stone with a gum arabic solution and rolled on greasy ink that would be accepted by the greasy image lines but not by the dampened non-image areas. He then pressed a sheet of paper over the surface to print the image.

Lithography is a process that lends itself more to printing the smooth tones of pictures than it does printing from the raised metal surfaces of Gutenberg's process, which came to be called letterpress printing. It solved some of the problems related to printing from raised metal type, woodcuts, and copperplate engravings. Lithography not only simplified the reproduction of multicolor pictures and improved tonal gradations, but it also made it easier to combine pictures and type on one plate or image carrier. Additionally, lithography provided an efficient means of printing complex, multi-character alphabets, such as those used in Asia.

Lithography eventually became the predominant process for nearly all printing applications. It still dominates today, using sophisticated printing presses using thin printing plates made of aluminum and other flexible metal image carriers that replaced stones in the 1930s and 1940s.

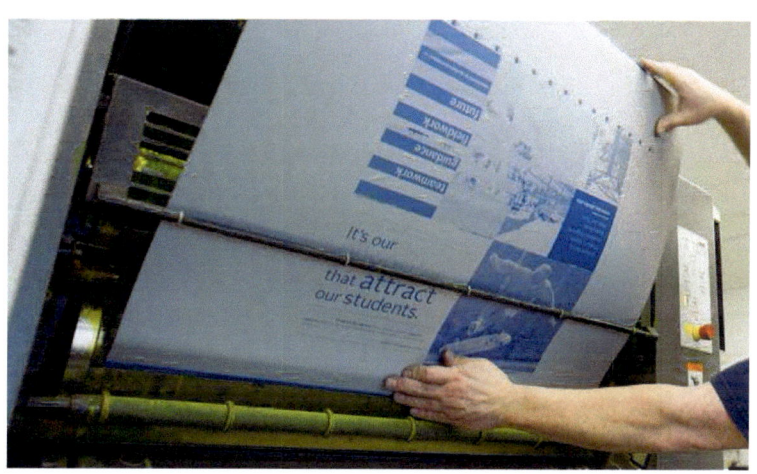

An offset lithographic printing plate.

Printer Ira Rubel replaced the transfer sheet with rubber "offset blankets" in the early 1900s, and the lithographic process became known as "offset lithography."

This became the dominant printing technology of the 20th Century, producing a steady decline in letterpress printing. Offset's dominance remained unchallenged until the advent of digital presses.

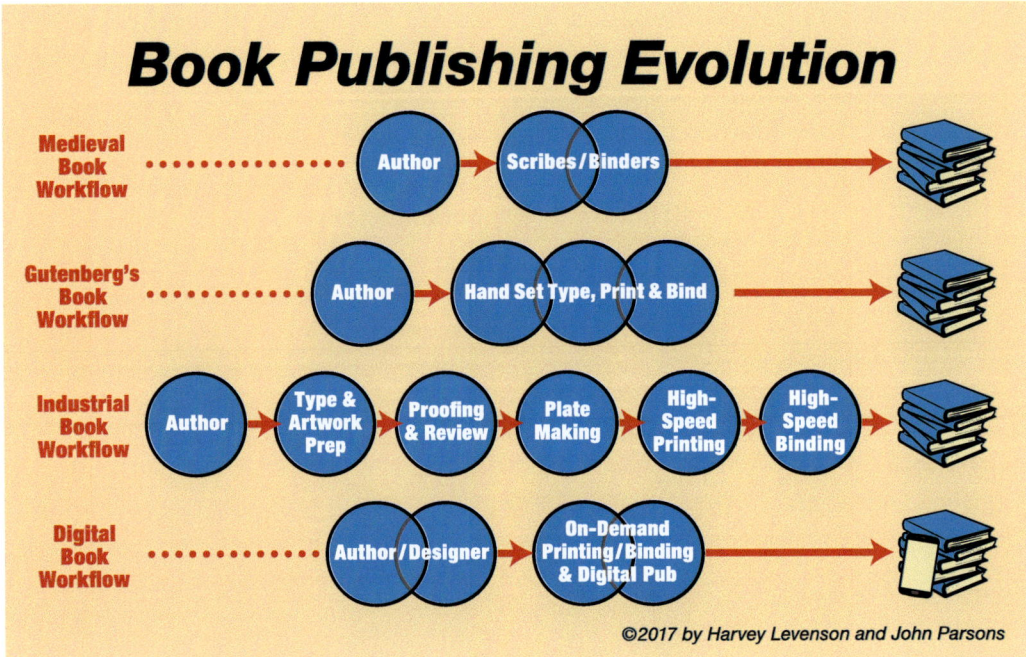

Publishing has evolved dramatically since the days of medieval scribes. Gutenberg instituted a change toward mechanized mass production, which grew into a complex sequence of industrialized steps. The digital revolution then combined steps and roles into a shorter, faster process.

Printing in the Electronic Era

The 20h Century saw the invention of many technologies that are essential to the graphic communication industry today. In the realm of printing on paper or other substrates, there were many steps that led to today's "digital presses," which compete in many ways with now-traditional offset.

Innovations in integrated circuits, transistors, and microprocessors have dramatically influenced the size, speed, and capacity of computers. These innovations have improved the printing industry in two significant stages. The first is that nearly every piece of printing equipment that has been developed since the mid-1980s is driven by a microprocessor.

The second is that the miniaturization of and cost reductions in microprocessor-driven technology have given nearly any person access to printing technology. This is most obvious in desktop publishing technology, where image generation tasks can be performed for under $1,000 that cost hundreds of thousands of dollars a decade earlier. Through such accessibility, the author of print can now also be the producer of print.

The second stage of the "author as producer" trend is the rapidly developing availability of high-quality, high-speed printing systems for mass production of text and pictorial images produced on the "desktop." The roots of this second stage were planted in 1938 when Chesley Carlson invented a process that helped make possible the office copier, a "printing plant" in the office. Carlson's invention became known as xerography, or "dry writing," and was accomplished by reflecting the image of an original document from a mirror onto an electrically charged selenium drum. A copy was created as special dry-ink powder (toner), attracted to the dark image on the drum, was transferred and then fused to paper.

Carlson's invention was an anomaly for printing systems of his time but represented a paradigm for printing technology concepts for the 21st Century. Sometimes referred to as "electrophotography," Carlson's xerography has become increasingly important in "on-demand" printing, and is the basis for Electrophotographic or EP printing (described elsewhere in the book). Other 20th Century developments include thermal and "piezoelectric" inkjet printing—also described elsewhere. These new approaches, combined with sophisticated bindery and finishing attachments have made such systems competitive with traditional lithographic presses.

The origins of desktop publishing resemble the days of medieval scribes and students of that time, who had to be paleographer, editor, and publisher of the authors they read. A manuscript book was costly; the simplest way of obtaining books was for the teacher to dictate the texts to the students, i.e., "desktop publishing." Such a commercial venture on the part of students who wrote and sold books assured teachers large audiences and, in some cases, substantial revenue.

Through the first half of the 20th Century, the competing technologies of radio and television expanded methods of communications. However, printing continued to grow. Electronics brought to printing capabilities that enhanced quality and the speed in which printing could be produced and distributed. In some cases, print moved to computer screens, but books, magazines, newspapers, manuals, and packaging all continued to grow in demand. In fact, desktop publishing has made printing more affordable and more available. Using computers, today's printer can create complex layouts, change colors, and manipulate images in a matter of minutes.

The advent of desktop publishing has broadened the scope of the graphic arts

Early Macintosh computers popularized the "What You See Is What You Get" or WYSIWYG approach to page layout and design, known as Desktop Publishing. (Apple Computer)

Introduction to Graphic Communication | 47

to include not only traditional printing establishments, but publishers, designers, advertising agencies, print buyers, service bureaus, print brokers, or any individual who desires to produce something as simple as a one-color letterhead or a complex color brochure. Desktop publishing has allowed the author and originator of an idea to become the producer of print media.

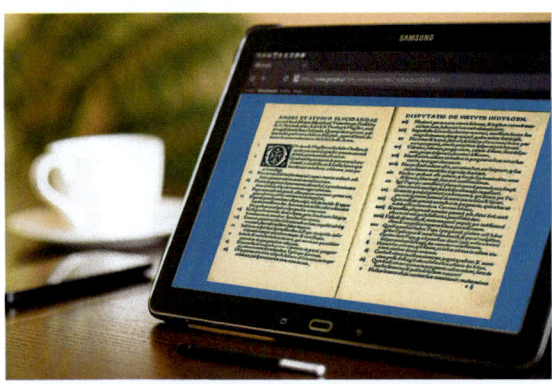

The printing of Martin Luther's 95 Theses was a result of historic changes in printing and publishing. The work itself was also a catalyst for social and political change. The same dynamic holds true for today's graphic communication technology.

On Beyond Paper

In parallel with new ways to put color on paper (or other materials), new technologies for manipulating and transmitting data, separating color images, and composing pages presaged radical changes, not only in the way we think of traditional printing, but also how we publish information without the use of paper. For much of that, we can credit (or blame) the Internet.

The process of digital, multimedia publishing resembles the disruption that followed Ts'ai Lun, Gutenberg, and Senefelder—but at an extremely accelerated rate. The mass reproduction of text and images on a sheet of paper closely mirrors that of doing so on a screen. The difference is the factg that the content is instantly changeable. We have only begun to cope with the implications of managing such volatile content.

Summary

Digital media, like its older and slower printed counterpart, is as likely to affect human society as Gutenberg's invention ever was. Graphic communication, on paper or on screen, will always be a catalyst for change.

Printing developed around the world because of technology transfer caused by changes in transportation, commerce, and the outcomes of war. However, the reverse is also true. Historic, often violent shifts in civilization are often triggered by changes in printing.

The invention of printing marks a turning point in the history of civilization. It has changed views about literary art and style and modified the psychological processes by which words are used for communication. The invention of printing broadcast the printed language and gave to print a degree of authority that it has never lost. Printing was the first public address system and soon became a new form of expression. It is an extension of human evolution and the innate need for humans to recall, keep records, and communicate.

3
Technological Transitions

Technological Transitions

Chapter Preview

The Impact of Rapid Change

From Macro to Micro

Lessons from History

The User Experience

Implementing New Technology

Overview

Geoffrey Moore's 1991 book, *Crossing the Chasm*, offered an insightful look at technology marketing. He wrote that the outlook and reasoning of enthusiastic early adopters differed widely from the vast majority of business decision makers.

This is especially true in the world of print, an industry with extraordinarily slim profit margins and enormous capital equipment costs. Making a bad decision on a multi-million-dollar press can doom a printing company. However, a good decision at the right time can mean remarkable success. With print-related technology changing on an almost daily basis, knowing when to move is critical.

This chapter takes a closer look at the dynamics of technology change in the world of print. It covers the overall nature of change, and how to evaluate and implement new technology successfully—mindful of its long-term business benefits, not just its "hype" factor. As is the case with most industries, novelty does not automatically guarantee success.

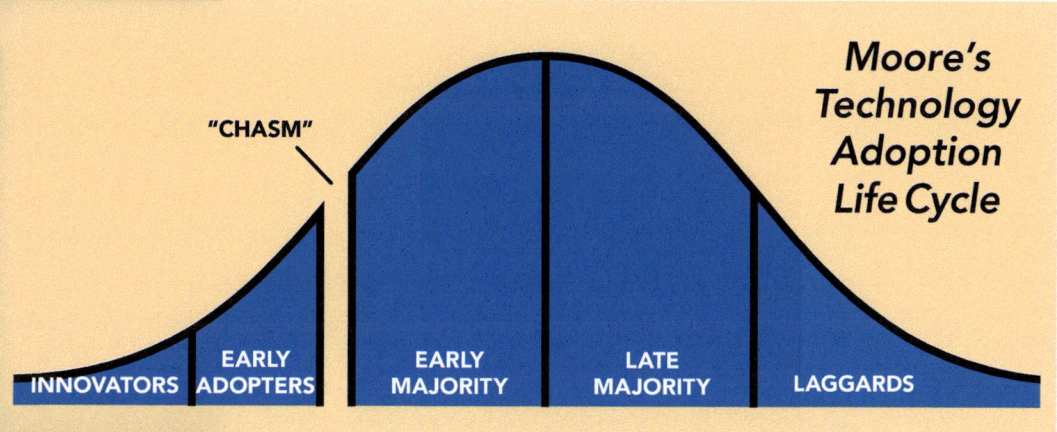

The Impact of Rapid Change

The history of graphic communication shows that the development of new technologies accelerates at a faster rate with each year. At the same time, the technology becomes more complex and versatile. Consider the following:

It was 430 years from Gutenberg inventing the process of duplicating movable type in 1456 to Mergenthaler's invention of the Linotype machine in 1886. However, it was only 68 years between the introduction of the Linotype and the invention of phototypesetting by the Harris Corporation in 1954. As with other industries, the rate of technology innovation in graphic communication has increased exponentially.

> **The rate of innovation in graphic communication has increased exponentially**

This technological transition time compression trend became evident in the graphic arts in the late 1960s. Dow Jones' use of satellites to transmit *Wall Street Journal* production data to multiple sites in 1968 followed the invention of phototypesetting by only fourteen years. Computer graphics on cathode-ray tubes took hold six years later in 1974.

In 1977, the Mitsushita Corporation demonstrated the potential of integrated media, building an inkjet printing system into a television set—to print out anything on its screen in full color. This was the forerunner of today's computer monitors, interfaced with printers for the production of desktop color.

Only two years later, in 1979, came the first practical combination of multiple graphic arts technologies from an Israeli firm, Scitex. It demonstrated how nearly all prepress functions (page layout, retouching, typesetting, graphic arts photography, and image assembly) could be incorporated in one integrated system.

Called the Response 300, this Scitex system planted the seed that led to the demise of traditional prepress departments in the graphic arts.

Micro to Macro

By 1980, transitions in graphic communication occurred so rapidly that it became impossible to use single developments to illustrate the rate of change. It was necessary to describe change in macro-technological development terms—occurring over a span of years. For example, from 1980 through 1985, the U.S. government deregulated telecommunication, which opened the airwaves for broad commercial applications of communication media and provided opportunities for the growth of digital printing and the Internet.

The birth, growth, and maturity of low-cost desktop publishing systems, from 1985–1990, were micro-technological developments. Combined with the macro-technologies of satellite transmission and Internet connectivity, it fostered subsequent macro-technology "waves," such as multimedia (1990–1995),

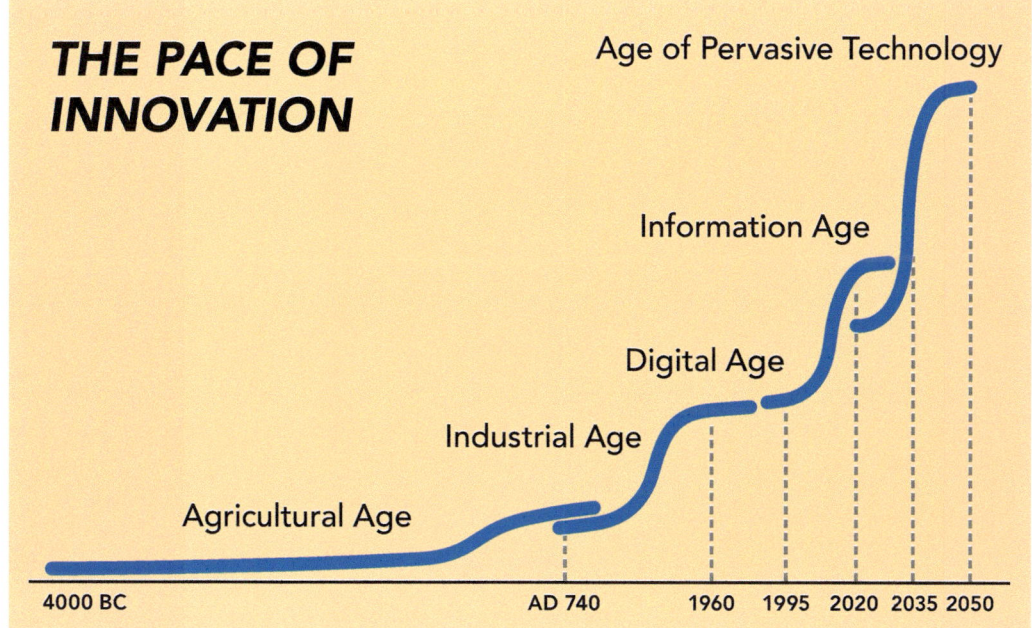

The pace of innovation accelerates at a point in every technology phase. (Courtesy of Wipro Ltd.)

on-demand digital color printing and web-to-print (1995–2005). A major macro-technology trend, cloud-based computing, has had enormous impact on almost every print-related system, from variable data and multi-channel publishing to remote collaboration and approval.

This trend will only increase. The early 21st Century marked not only the decline and consolidation of traditional processes; it also heralded the emergence of new combinations of print, digital, and even virtual reality-based graphic communication.

Although the rate of technological transition has accelerated over the last 500 years, the need to understand the impact of change remains the same. All new technologies have to go through the product development and marketing steps necessary for technology to successfully progress from concept to market demand and acceptance.

Lessons from History

The following principles by the late Jim Wilkins, former director of Printing Industries of America, address issues of how technologies change. They also suggest strategies for effective implementation and marketing of technology in the graphic communication industry.

Technologies tend to converge. One technology borrows from others. No single science has all the answers to solving the changing needs of society. For example, in graphic communication, the sciences of optics and chemical emulsions converged to form the technology of photography. Design tasks and digital computers converged to form integrated, multi-function systems. Digital printing converged with the Internet to create Web-enabled printing.

Not all such combinations are as significant as these. But very few graphic communication systems exist that do not borrow from (or even cannibalize) successful technologies of the past.

> Customers expect high-quality, low-price products delivered quickly. But this can no longer be provided using older technology.

Change is typically evolutionary, but is increasingly revolutionary. By the end of the 1990s, there were very few graphic arts technologies that had changed the posture of the entire industry within a five-year period. Previously, the time interval between invention and widespread implementation was ten years or longer. It took about twenty years for the Linotype machine to have an impact on the printing industry, but only ten years for phototypesetting to have a comparable impact.

Since the 1990s, the time it takes for a new technology to fundamentally change the industry has been greatly reduced. Broadband Internet communication and digital photography are two prime examples affecting society in general and graphic communication in particular.

Technology tends to expand existing markets and create new ones. Contrary to popular belief, technological innovation does not eliminate markets. Instead, it improves the ability to serve and expand existing markets and provides opportunities to develop new ones. For example, the development of the rotary offset press increased the speed with which printing could be produced and provided opportunities for printers to develop commercial markets in areas such as advertising.

More recently, short-run digital and variable-data printing has significantly disrupted the *status quo*. However, for the most part, it has not cannibalized printing jobs designed for traditional presses. Rather, it has provided opportunities to service the short-run digital and traditional printing needs of clients.

Technology and practical needs share a mutual attraction. Technological change is rarely as disruptive as expected. For example, the advent of radio did not displace newspapers, nor did the advent of television displace radio or cinema. Similarly, the Web did not displace commercial printing but provided new opportunities for print to support Web communication and vice versa. Although service businesses dependent on older technologies often suffer during disruptive change, new opportunities for growth are always present.

Results are measured in practical, human terms. Technological change can be postponed or slowed down, but it cannot be stopped. Eventually it breaks through—to satisfy basic human needs. In the mid-1950s, Linotype-centric unions forestalled the introduction of phototypesetting, but eventually the need for efficiency and speed overpowered the desire to maintain the status quo. Similar resistance to desktop computers, email, and mobile phones eventually succumbed to the practical results of widespread use.

In the give and take between people and technology, people prevail. Technology must benefit people because they create and control technology to satisfy human needs. People will not revert to old technology to fulfill practical needs once new technology is adopted. For example, many designers resisted the idea that original art could be done on a computer. Yet once the new electronic medium was understood and perfected, many artists no longer resort to using canvas or illustration board.

Customers' expectations have also changed with new technologies—they want high-quality, low-price products delivered quickly. But this can no longer be provided using older technology. Once the new approach meets the basic, human need, change is irreversible.

The User Experience

To be successful in the marketplace, technological innovation must address more than just technology. It must address the human interface to all technology. Questions that help determine if a technological transition is ready to take place include:

- Does present technology no longer serve a need?
- Are people ready for a new technology to the point where they will invest in it?
- Will a new technology be embraced as a means to make work easier, but not less important, and as a way to better service market demands?

The most successful new technologies are those that present a proven paradigm shift in the form of a device or concept that helps solve old problems in a new and better way, or the technology aids in understanding ideas, systems, and behaviors.

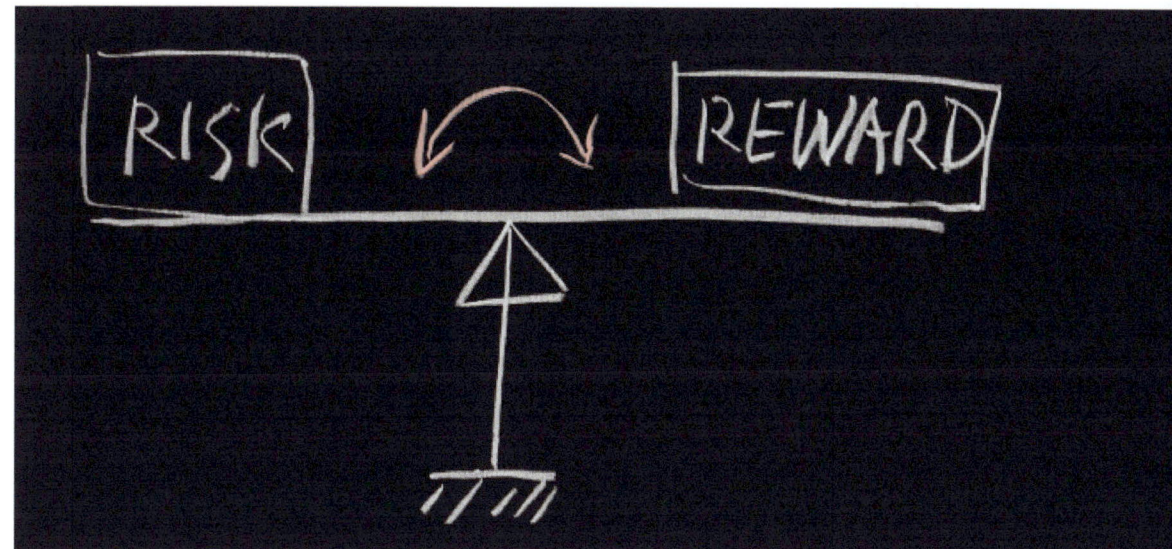

Introduction to Graphic Communication | 55

Implementing New Technology

According to Dorothy Leonard-Barton of the Harvard Business School and William Kraus of the General Electric Corporation, there is a natural resistance to technological change. This stems from a gap between the value of technology and the ability to put it to work effectively.

In many cases, a large investment may have been made in a new technology, but the technology is subsequently not used to its maximum potential. Examples in the graphic communication industry include the acquisition of high-end, integrated systems and very expensive and powerful computers. Digital printing presses too often fall into this category, since they are a major capital investment, but can often be under-utilized.

> *All cost factors must be taken into consideration when evaluating a new printing technology.*

A company may acquire a high-end system and use it for only one or two applications. This occurs when the company is attracted to the glamour or "magic," of the system, with every intention of using its capabilities to expand markets and attract new ones. However, this rarely happens as planned. Companies must handle the day-to-day routines of serving its traditional client needs, and the task of new market development can be neglected.

Sometimes the problem is resistance from those who are expected to operate the new technology, but are not properly prepared for it.

Yet another problem that graphic communication companies face is overcapitalization in order to bring in promising new technology. This makes sense if the company is expanding its markets and generates additional revenues from the technology. However, this does not always happen. Companies that make large investments solely to attract and keep customers may soon find themselves out of business.

There is a distance between technological promise and achievement. Does technology do what its advertising claims it does? Advertising tends to exaggerate the ease with which technology can be maximized.

Advertisements seldom address the cost of operating technology, the life of components, the need for maintenance contracts, the cost of supplies. These must all be taken into consideration when determining whether a technology will be cost effective. For example, nearly all major electronic printing equipment requires the purchase of a service contract. This is rarely noted in product advertisements or brochures—nor is the cost of consumable supplies (such as inks and toners) or replacement parts.

Leonard-Barton and Kraus list six requirements for implementing new technology successfully. Later chapters will outline the major areas of print technology, which are constantly evolving and changing. No matter what technology is involved, OEMs, vendors, service providers, and their customers should remain acutely aware of these principles.

Dual role. Because the user of new technology is not always ready to use it, the developer or manufacturer of the technology must implement it at the plant where it is being installed. With very sophisticated technology such as large digital printing presses, the manufacturer will often put a technician in the plant for an extended period of time. It could be months before plant personnel are ready to operate the technology unassisted.

Marketing perspective. An intelligent approach for marketing and implementing any new technology is to involve users in its design phase, or at least in the planning process. Additionally, prior to installation, new technology must be marketed to operators. Ideally, they should be convinced that the technology is most suited to meeting the goals of the company, and operators should be included in the decision to acquire it.

For example, a major printing company in Canada sends operators to non-competitive plants in Canada and the United States to see a particular technology at work and how peers are using it. Another company sends operators to expositions where new technology is being exhibited and to manufacturing plants where it is being built with the hope that the operators will recommend purchase.

The theory is that if operators endorse it, there is a greater likelihood that the technology will operate successfully, and will result in greater productivity and profitably.

Framework for information. Ideally, one person should coordinate the work of gathering all of the information needed for the implementation of new technology. That person should observe current job routines and workflows and identify bottlenecks that need to be eliminated. An analysis also should be conducted to ensure that the new technology would not create bottlenecks, as when production in one department increases to the point where the next department will be unable to handle the increased volume.

> *Success with a technology in one plant does not guarantee success in another.*

This latter problem occurred in the graphic communication industry with the automation of prepress in the early 1980s. Graphic arts photography via scanning, image assembly, and platemaking started to occur so rapidly that press departments could not handle the products of prepress fast enough and bottlenecks occurred. Hence, the next focus of automation was in the press department, and the late 1980s and early 1990s saw tremendous improvements in press speeds through the introduction of electronic press controls. By the early 1990s, sheetfed and web press speed nearly doubled and wiped-out the bottleneck between prepress and press. However, the elimination of this bottleneck caused another one in binding and finishing departments, where printed sheets or printed web rolls had to be slit, scored, folded, collated, trimmed, and so on. Thus, in the mid-1990s, great improvements in binding and finishing speeds were developed, particularly in the form of online and integrated press and finishing functions, and this reduced bindery bottlenecks.

> New technology can sometimes change interdepartmental relationahips and dependencies.

Attention also must be paid to the sections of work that require user decisions about tools and materials; safety and health issues apply here. New technology sometimes raises legal issues related to employee safety and environmental protection. Someone must take responsibility to ensure that all Occupational Safety and Health Act (OSHA) and Environmental Protection Agency (EPA) regulations are being met with the implementation of new technology.

In the printing industry, exposure to chemicals and volatile gases, and waste disposal are grave concerns. Some technology requires chemicals that must be appropriately shielded from machine operators, and such protections are typically very costly. Disposal of unused inks, fountain solutions, photographic chemicals, plates, and related waste also must be monitored. Without appropriate protections in these areas, successful implementation and operation of new technologies may not be possible.

A common occurrence in the graphic communication industry is a case where technology works at one company with no problems whatsoever, but a second company using the same, generally reliable technology has ongoing difficulty with implementation and operation.

For seemingly indefinable reasons, the employees of one company, for instance, will swear by the superiority of a particular printing plate, but another company doing the same type of work on a similar press cannot run that brand of plate without encountering problems. The same situation happens with supplies such as inks, fountain solutions, and so on; and to equipment such as scanners, exposure units, computer-to-plate devices, and presses. Hence, if the technology itself is reliable, there must be something else causing inconsistency of performance from one company to the next. Employee attitudes about the technology, how they are introduced to it, their personal relationships with equipment and supply vendors, and what they hear from others all play a role in the motivation to implement and operate technology successfully.

The person responsible for coordinating the work of implementing technology should also investigate how manufacturing processes within the company relate to each other and determine the extent to which machine operators are dependent on materials, personnel, maintenance, and so on.

New technology sometimes changes technical relationships and dependencies between departments or cost centers. Such changes could provide improvements or hindrances, and companies replacing old technology with new should sufficiently study the impact on workflow to avoid surprises. For example, supplies have to be at the right place at the right time to continue a smooth flow of production. A question to ask before implementing technology is: Will changes in technology impact supply availability?

The graphic arts industry is reliant on systems that ensure production flow in a way that minimizes downtime, waste, and labor cost. Industry segments such as commercial printing are extremely competitive and rely on low profit margins to obtain and retain customers. The smooth and expeditious running of most printing plants relies on a rigidly coordinated effort between all cost centers of production. For example, the efficiency and quality of work produced in prepress influences the ease with which work can be performed properly in press departments. Likewise, the quality of work performed in press departments influences how efficiently work can be produced in the binding and finishing department.

Specifically, in the traditional flow of printing production there are six cost centers: art and copy preparation; graphic arts photography (though nearly obsolete), including scanning; image assembly; platemaking; presswork; and binding and finishing. While there may be fewer cost centers in highly automated companies, the concept still applies. Typically, the earlier stages of production represent lower cost centers. In other words, it cost less to perform tasks in art and copy preparation departments than it did in graphic arts photography, image assembly, or platemaking. This has nothing to do with the abilities or qualifications of the people who worked in these areas, but related to cost of labor (number of people working in these areas) and the cost of technology required in each cost center.

Art and copy preparation departments, even when heavily automated with computer systems, required fewer people and less expensive equipment than other prepress departments of printing plants. Likewise, press departments and binderies are relatively labor- and capital-intensive, and they require more expensive equipment and supplies. Printing presses often cost millions of dollars, and paper—the costliest disposable commodity in printing—typically represents a very large percentage of the cost of printing.

In other words, if it cost a client $2 million to purchase a printing job, it is not unusual for approximately $1 million of

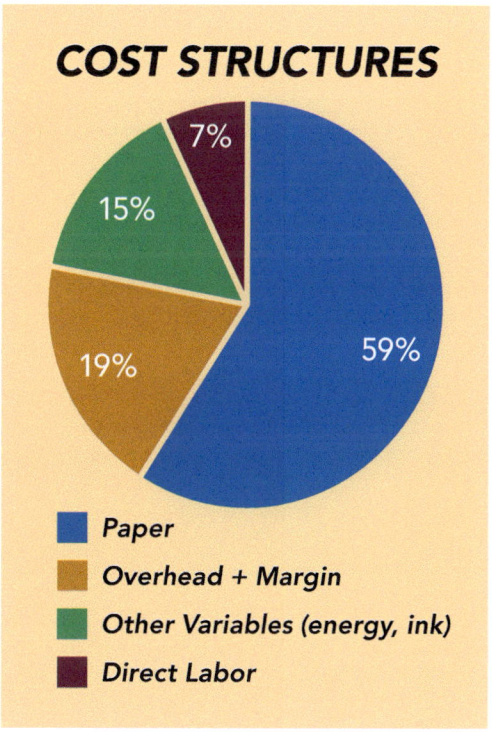

The economic realities of print include fixed costs, as shown in this breakdown for a simple, printed circular.

Introduction to Graphic Communication | 59

the cost to go toward paper. With this reverse pyramid of cost, the "system concept" applied to printing production teaches that work should be organized in such a way that as much work as possible is performed in the lower cost centers, thus minimizing the time it takes to produce work in higher cost centers.

One important goal is to avoid backward movement, i.e., having to return work to earlier cost centers. This could happen, for instance, with poor imposition, or improperly prepared plates that are not discovered until a job is already on press. Another example is when files are not prepared properly in a company's IT department.

Such breaks in the flow of production often mean the difference between profit and loss in the printing industry. One of the costliest interferences in production flow is when paper becomes unstable and stretches or shrinks due to relative humidity (RH) changes around the press and in the paper; this will adversely affect color registration and paper movement through presses.

> Too often, printing OEMs develop new technology without addressing needs of individual users.

Multiple internal markets. The higher the organizational level at which managers define a problem or a need, the greater the probability of successful implementation of new technology. When top management shows genuine concern that a new technology must solve certain defined problems and that the technology must be made available within a defined but reasonable cost range, developers will focus on meeting these production and bottom-line needs.

All too often, for instance, printing equipment manufacturers develop new technology without addressing the focused needs of individual potential users. The problems that the new technology addresses—such as faster speeds, higher quality, smaller "footprints," safer conditions, and so on—are broadly defined by the developer and marketed to companies that are experiencing those problems.

On the other hand, a company seeking new technology to solve problems specific only to itself is best served when top management approaches developers with the request to develop the technology or modify existing technology. When top management takes on such an exploratory role, this is evidence of the seriousness of the need for and the likelihood of contracting for the technology prior to creating it.

Promotion versus hype. Overselling a system can be dangerous to successful implementation and profitability. There is a gap between the perception of what a technology promises to do and the reality of what it can actually do that must be closed. People develop negative attitudes toward technologies that do not meet original expectations.

Hyperbole is not a new concept when it comes to technology in general.

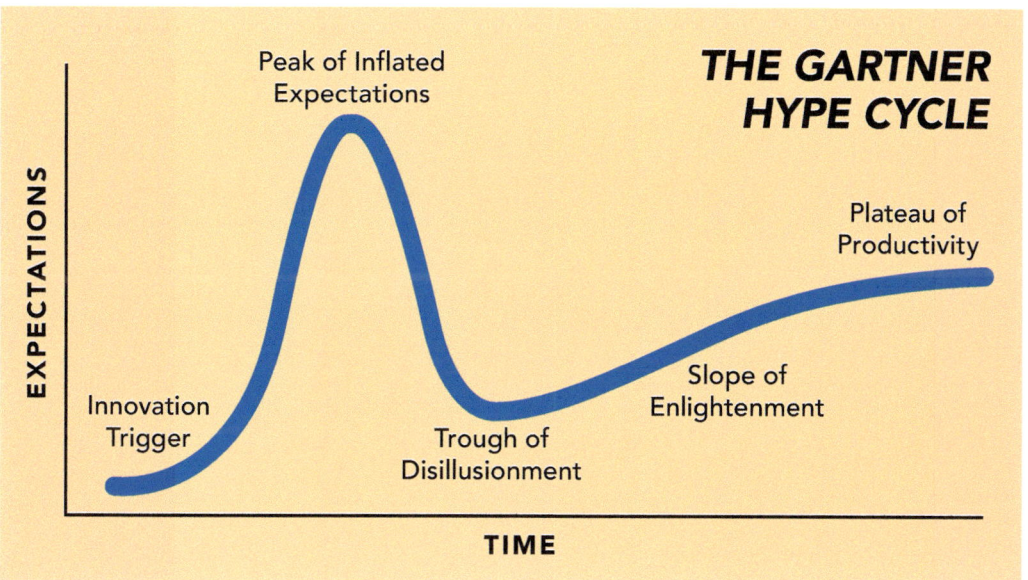

Breakthrough innovations have always generated initial public excitement, followed by severe disillusionment, and eventually a gradual increase in adoption and best practices. This has become known as the "hype cycle," as developed by the technology research firm Gartner Group.

Mapping specific graphic communication technologies to exact points in the Hype Cycle is outside the scope of this book. However, a recent example is worth noting. In the late 1990s, the industry "discovered" the idea that print jobs could be specified and ordered online. Originally dubbed "print e-commerce," this became known as Web-to-Print (WtP), and is covered in Chapter 12.

Just before the dot-com collapse in 2000, there were over 140 companies in this space—many with outlandish claims and enormous venture capital backing. After the crash, a much smaller group of service companies and developers created second- and third-generation WtP systems, and have begun to create viable best practices.

Risky site—safe innovation. A pilot site helps to iron out potential implementation problems and raises questions that must be answered before going to market. Ideally, testing should occur at a subnormal performance site, to answer real-world questions about implementation. This is not always the done. Very often, however, new technology is tested in a beta site representing pristine and even antiseptic conditions, e.g., a clean environment, excellent lighting conditions, and well-trained operators. However, this tells nothing about how the new technology will work under typical operating conditions. Hence, the recommendation is that, for example, beta testing a new printing press in a plant having relatively poor environmental conditions and minimally trained operators will tell more about the functioning of the new technology than would testing in an ideal environment. If the problems discovered under poor conditions can be resolved, then the technology is likely to work well in any environment.

4

Print Industry Segments

Print Industry Segments

Chapter Preview

Commercial Printers
On-Demand Printers
Specialty Printers
Publication Printiers
Financial/Legal Printers
Direct Mail
Packaging
In-Plant Printers
Service Providers
Print Brokers

Overview

The graphic communication industry is composed of many different types of businesses, each offering a selection of print-related services.

In some cases, these services also include online elements, such as email campaigns and customized landing pages that accompany a direct mailing. In other cases, the type of printing process is dictated by the immediate customer need, as in packaging. Some companies also attempt more than one type of service.

Each industry segment grapples with its own set of challenges, and strives to provide a distinctive and valuable service to graphic communicators.

This chapter is a general overview of the major company types engaged in printing, publishing, packaging, and related areas, along with an overview of their business trajectories.

A Complex Service Industry

There are many different types of companies or business units in the printing and graphic communication industry. These classifications can be problematic. For example, the U.S. government identifies a number of print-related business types according to its North American Industry Classification System (NAICS) number. However, due to rapid technological change and the tendency of companies to diversify their services, government classifications do not always apply.

> *Of the 36 print-related business types, some are on the brink of obsolescence, while others define the future of the industry.*

There are 36 industry segments related to printing. These range from the longstanding—some on the brink of obsolescence—to the latest and most vibrant segments that define the present and future of the industry.

The 36 segments span four categories: content creators; service providers; developers, dealers, and distributors; and those involved in education and professional development. Because this book focuses mainly on print service providers, the following is a summary of the types of companies that mass-produce printed content prior to distribution.

Commercial Printing

As of 2017, there were approximately 26,000 commercial printing establishments in the United States. This number has been dropping steadily due to a continuous flow of mergers, consolidations, and businesses terminating operations. This segment performs general printing of everything from simple flyers to complex four-color and spot-color printing, most often on paper substrates. Most print runs are relatively short, and getting shorter as digital printing displaces conventional offset lithography. Over 80% of all commercial printing companies employ fewer than twenty employees.

The commercial printing segment is highly competitive and relies on high total volume, with low profit markup, to sustain itself. Commercial printers typically provide a range of design and prepress services, and offer a wide selection of paper, ink, finishing, and fulfillment options.

With an emphasis on custom service, this segment tends to focus on local businesses, although larger commercial printers can offer cost savings to regional and national companies. As is typical of most industry segments, the rapidly growing area of commercial printing is in color reproduction and digital printing.

Because of the nearly infinite variety of printable items, print processes, and order complexity, commercial printers typically rely on a network of printing and fulfillment partners, and use online ordering to streamline the process.

APPOINTMENT CARDS, VOUCHERS, WRAPPING PAPER, CALENDARS, DIRECTORIES & YELLOW PAGES, BOOK JACKETS, SCRATCH CARDS, SCHEDULES, BOXES, FLYERS, POSTCARDS, TENT CARDS, WALLPAPER, BUSINESS CARDS, CATALOGS, REPLY CARDS, LETTERHEAD, SLEEVES, **BROCHURES**, FORMS, FLAGS, BAGS, POINT-OF-PURCHASE, INSERTS, ANNUAL REPORTS, BOOKS AND BOOKLETS, SIGNAGE, LABELS, MANUALS/GUIDES, GARMENT TAGS, JOURNALS, POSTERS, EMPLOYEE BENEFIT KITS, PRICE/SALE TAGS, BINDERS, LEAFLETS, PROMO CARDS, STICKERS, NOTEPADS, COMPANY REPORTS, CD INSERTS, BOOKMARKS, FRIDGE MAGNETS, MENUS, INVITATIONS, POCKET FOLDERS

The decline in the number of printing establishments is slowing and will stabilize at about 22,000 industry firms. Print is still required for many products, notably packaging, that cannot be substituted by electronic methods. For the printing companies that remain, the business outlook is good. There will still be a need for large, complex printing projects.

On-Demand (Quick) Printing

Arguably a variant of commercial printing is on-demand or quick printing. Originally known as copy shops, these are typically walk-in, storefront service companies, offering quick turnaround on relatively simple and low-volume print jobs. Although such shops once used one- and sometimes two-color offset presses (called duplicator presses), quick printers today almost all use digital printers that also serve as photocopiers.

On-demand printing companies often works closely with individual designers, providing access to desktop design computers and scanners for a fee. Output and binding options are simple in nature, with a marked increase in color output on uncoated paper. Mailing services are frequently offered, as are basic web-to-print online storefronts for submitting print jobs.

In 2017, there were approximately 4,900 on-demand or quick printing locations in the U.S. Many of these are individually-owned companies, but many are part of franchises such as AlphaGraphics and FedEx Office (formerly Kinkos) or are co-located in stores such as OfficeMax or The UPS Store.

There are four levels of print: do it yourself on a home printer, go to a local quick printer, go to a large commercial printer, or order online. The use of any of these options depends of the size and complexity of the job and the schedule. There will always be a need for "longer short runs" that can be produced by a local quick printing operation.

Specialty Printing

Often, a particular printing application—with its own, unique substrate, ink, and equipment requirements—is beyond the capabilities of a general commercial printer. Specialty printing businesses can be of any size, and can be independent or part of a larger printing enterprise.

Specialty printing, also called functional or industrial printing, involves the use of substrates beyond paper. New flatbed inkjet printers can print on plastic, metal, textile, ceramic, wood, glass, carpeting, and other materials. This opens new markets in home decorating, building trades, apparel, engineering, fine art, and other emerging markets.

Specialty printing is emerging as one of the fastest growing areas of the graphic communication industry. This is due to the growth of variable data digital inkjet printing, and the ability of to print directly onto preformed products profitably in short runs. The Screen Printing Association changed its name to the Specialty Graphic Imaging Association (SGIA) to address the versatility that specialty printing now provides.

Sign printing. Traditionally printed via offset lithography or screen-printing, signs are increasingly the domain of wide-format digital inkjet printing. The finished product must be viewable and effective at longer distances—as opposed to publications and collateral, which are meant to be held and viewed closely. They must also be durable under a variety of indoor and outdoor conditions. For signs that cannot be printed as a single sheet, there must be an effective way to combine or "tile" individual sheets into a single piece, as in the case of billboards, vehicle wraps, and signage covering entire buildings.

Some kinds of signs are not printed at all, such as those made with neon lights, channel lettering, or laser-cut wood and other surfaces. When printing is employed, however, customers can choose from a vast array of substrates, particularly vinyl, transfer film, and specially-coated paper. Each of these requires specialized inks and equipment.

Sign printing is a complex and growing market segment, with about 3,100 companies in the U.S. in 2017. These companies typically work closely with designers and brand owners, not only to handle the many printing and finishing variables, but also to help manage color consistency under different printing and environmental conditions. Sign printing companies also frequently offer related products, such as indoor display systems.

Building wraps are a highly-specialized form of signage production.

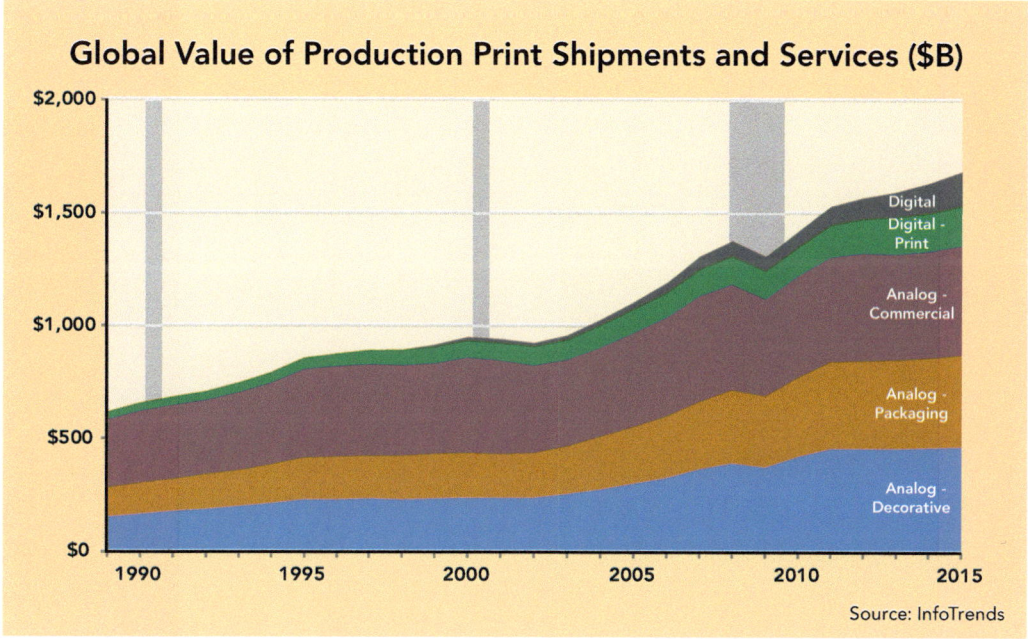

Except during periods of overall economic downturn (vertical gray bars), most print industry sectors are experiencing significant, albeit modest growth. (Data courtesy of Frank Romano and InfoTrends.)

Promotional item printing. A very specialized branch of this segment involves printing on non-paper substrates such as textiles and various plastic and metal objects used for promotional purposes. This can include a very wide range of items, from t-shirts to coffee cups and branded USB drives. Traditionally, this involved screen-printing, but increasingly involves digital printing.

This segment overlaps many others, including packaging (where printing on non-paper substrates is common) to textile printing in general.

Business forms and bank stationery printing. Business forms printers are rapidly disappearing due to software that allows such products to be produced on as desktop or laptop computer. However, such companies typically produced three main products: snap-out or unit set forms, computer or continuous forms, and specialty forms. Snap-outs, used for utility invoicing, include a mailing envelope, an invoice, and a return envelope all gathered in one unit. Computer and specialty forms, once an essential part of office infrastructure, are being supplanted by their digital counterparts, and thus comprise a declining portion of printed business forms products.

With the proliferation of electronic funds transfer systems and ATMs, bank stationery printing as a separate industry segment is being eliminated. Bank stationery products still in use, however. Checks and deposit/withdrawal forms are typically produced by this segment.

Introduction to Graphic Communication | 69

For many years the business forms segment was highly profitable, because it manufactured specialized products on standardized equipment with a relatively small variety of supplies. Its small format and narrow web presses were relatively simple and required low operator skill levels. Business forms products are typically produced on uncoated paper. Color printing for business forms has now become a standard.

While the use of printed business forms is on the decline, an interesting growth area is in printing personalized forms with information directed to the individual recipient. Many utility bills now typically carry personalized messages and advertising directed to the addressee. These mailed forms utilize variable-data printing equipment and sophisticated databases, as described in Chapter 8.

Other specialty printing segments. The highly-specialized greeting card printing segment is growing, despite the advent of Internet-based electronic greeting cards. It is also dominated by a small number of companies, with about 80 percent of all greeting cards in the U.S. produced by few companies, led by Hallmark. Greeting card printing involves an extremely wide variety of substrates, custom color inks, metallic pigments, laminations, and coatings. It also requires specialty processes such as embossing, die cutting, and holography.

The dominance of this market by Hallmark and others has resulted in one of the most demanding print color quality control and monitoring programs in the industry. A current trend is to manufacture regional or "demographic" cards that are identified with a particular community, city, or state, and personalized cards on which the consumer can add a personal message or a picture.

Another specialty printing segment—school yearbooks—resembles that of traditional book or magazine publishing (see below) but with an emphasis on emotional appeal over sophistication. Because of its seasonal nature, predictable demand, limited format and color options, and reliance on relatively unskilled or temporary labor, yearbook printing remains profitable.

Publication Printing

Printing is historically associated with publications, ranging from works with no advertising whatsoever (books and journals) to those supported by wholly or partially ads (newspapers and magazines). Arguably, printed pieces that are entirely supported by a commercial entity (catalogs and other direct mail) could also be

Publication printing is the industry segment most impacted by digital media.

classified as publications—at least from a print production standpoint.

Of all industry segments, publication printing is the most impacted by digital media. The economics of printing books, newspapers, and magazines—at high volumes for a mass audience—has been severely disrupted by online dissemination of news and information. Publishers are slowly adapting to this change—initially via digital replicas of their printed products—with mixed success.

Books. Book printing and publishing has been a slow growth area for many years. Strained school budgets and disruptions in learning technology have kept textbook purchases flat, despite growth enrollments. For fiction and nonfiction books, competition from digital and broadcast media has influenced the degree to which people read books for personal enjoyment. In the past, a growth area of the book printing segment was the production of manuals and technical books. However, that too has been impacted by digital versions which, in theory, are better able to adapt to rapid technological change.

The book industry is characterized by mergers and acquisitions among publishers and distributors. Today, the majority of trade books are published by only a few companies. The current "Big Five" publishers are Hachette, HarperCollins, Macmillan, Penguin Random House, and Simon and Schuster. However, as the major publishers consolidate, small, independent publishers and self-publishing businesses proliferate.

Book distribution is also dominated by a small number of very large companies

such as Ingram, Barnes & Noble (physical bookstores) and Amazon (online sales, print-on-demand, and eBooks).

A large portion of this industry segment involves black-and-white printing on uncoated paper, using sheetfed, perfecting presses. Color printing on coated stock is increasing for most book categories outside of trade fiction and nonfiction. While offset lithography is still significant, high-speed inkjet presses are increasing in many book printing operations. Paper for books has also improved significantly, with the efficient manufacturing and printability of recycled, acid-free, and alternative paper sources.

Technology shifts—especially supply chain automation—have driven much of the industry consolidation involving high-volume book production and distribution. However, technologies such as desktop publishing and on-demand printing have also made it possible for small publishers and self-publishers to succeed. Extremely short runs are now cost-effective, reducing the need for

Although still viable internationally, newspaper printing is significantly affected by the rise of digital media.

high-volume inventory and storage of books. Textbooks customized for individual or classroom use are now economically feasible, thanks to variable data printing on inkjet presses. This will also let publishers offer limited-edition books geared toward regional preferences.

Globalization of this industry segment will grow as the technology for transmitting, editing, assembling, and printing book content—from any place in the world—becomes more commonplace and accessible. Similar to Gutenberg's 13th Century invention, the digital publishing revolution will make books more accessible and versatile.

Newspapers. The segment is defined as that which prints on newsprint—a lower grade of wood pulp-based paper typically printed on roll-fed presses. This includes not only newspapers themselves, but also advertising inserts and special publications. Daily newspapers are on the decline in the United States, dropping from about 7,800 in 1990 to about 3,000 in 2017. However, weekly, community, and special interest newspapers continue to serve local and distinctive interests of individuals and groups. Some newspapers have experienced revitalization and growth resulting from mergers and acquisitions, and from interactions with other members (broadcast and digital) of an encompassing media conglomerate.

Originally the exclusive domain of letterpress, newspapers are now produced primarily using web offset or flexographic printing technology. (See Chapter 8.) The use of gravure printing, formerly for color advertising inserts and comics, is in sharp decline as circulations decline and as color quality for offset and flexographic printing has vastly improved.

The increased use of color for non-advertising sections of newspapers was spearheaded by Gannett's USA TODAY in the early 1980s. Since then, most major metropolitan and many community newspapers have color integrated throughout the publication, not limited to special sections. Technological advances for color reproduction include the development of "non-rub" inks—water-based flexographic inks resistant to flaking or scuffing.

Although print advertising is still the mainstay for newspapers, many are moving rapidly into digital applications, diversifying their services through the use of online databases. These supplement the printed versions of their products in editorial content and classified advertising, and more recently supporting mobile applications and online content syndication services.

Magazines and periodicals. For many years, the trend in magazine and periodical printing and publishing is toward specialty publications. General interest or news magazines have given way to those focused on fashion, health and fitness, sports, travel, entertainment, technology, home improvement/decorating, and an array of personal vocations and hobbies. These typically have smaller circulations than general interest magazines, and are more likely to be distributed by mail than via newsstands. (Specialized distributors such as Barnes & Noble can be the exception to this trend.) Regardless, the collective circulation of all magazines and periodicals remains vibrant in spite of postal rate increases.

A major characteristic of printed magazines is the use of color and higher-quality, typically coated paper. The tactile aesthetic of a printed magazine—and its reputation as a more relaxed, "lean-back" media experience—is often cited as a reason for its continued success, although digital and mobile editions and apps are used by many magazine publishers.

Magazine production today is electronic in the preparation of layouts, pagination, and image assembly. This industry segment will continue to grow as a result of the increased availability of low-cost and sophisticated desktop publishing systems. Magazine publishers also increasingly use technology that allows for demographically-segmented printing and distribution. This lets magazines provide advertising and even editorial content of interest to readers of a particular geographic region. Some publishers offer split runs and deliver by zip codes, with selective binding used for personalized advertising.

Catalogs and Directories. Many professionals object to considering these as publications at all. From an editorial and economic standpoint, they are correct, although some consumer magazines have blurred the line between editorial and advertising to the extent that they resemble catalogs.

Consumer catalogs, once very large, multi-page publications are now much smaller and more specialized. For retailers seeking a sensory, haptic connection with the consumer, their catalogs require a greater focus on color and high-quality paper.

Introduction to Graphic Communication | 73

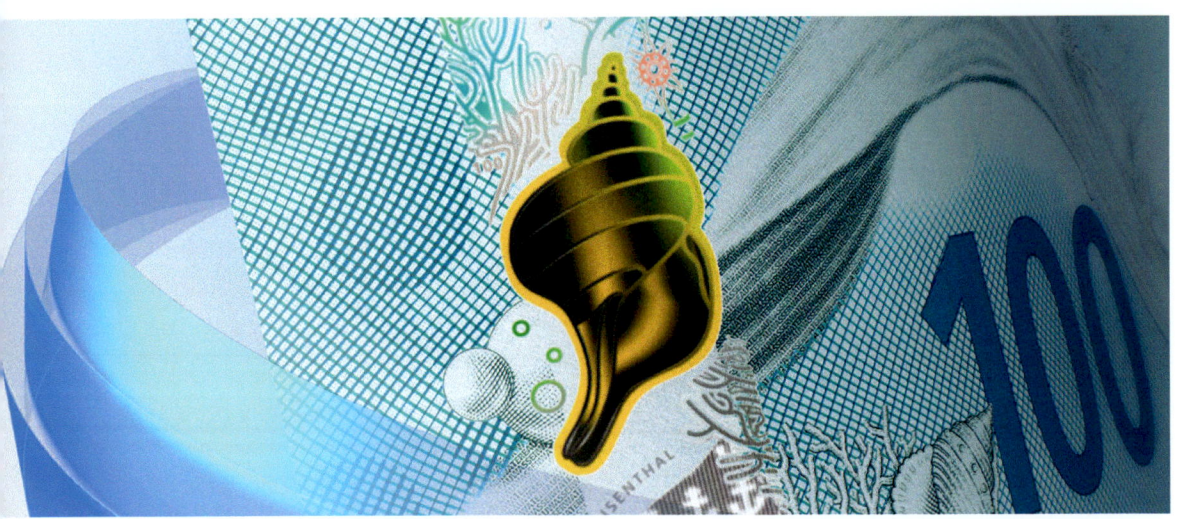

Security printing, including currency, banking and financial documents, and lottery tickets, include many techniques designed to prevent counterfeiting, including metallic elements, special paper, and holograms.

From a purely production and distribution standpoint, however, catalogs are remarkably similar to magazines, and are very often printed with the same equipment. Both use high-resolution color printed on coated stock, and use the same binding styles. Both also use digital printing to supplement web offset production, mainly to print customized mailing information and other variable data elements. Catalogs are distributed primarily through the postal system, although some have found their way to the newsstand.

The use of and interest in printed directories, such as telephone books (white and yellow pages) has decreased significantly due to online offerings of similar information. The speed of locating listings online, and the ability to activate a phone call or a web page with a click—often integrated with mobile ads and map applications—are features that printed directories cannot offer.

Even with electronic publications, print is holding its own. On-demand printing has reduced costs for book publishers and printed books are growing. Magazines and newspapers are offering electronic versions but still publishing smaller print editions. Users like the fact that they can save and share printed versions of their publications without the licensing and technology overhead that digital versions require.

Financial and Legal Printing

This segment produces materials related to monetary and contractual situations of all kinds. The emphasis for printed products is on security and fraud prevention, and include stock certificates, lottery tickets, traveler's checks, bonds, leases, SEC filings, currency, and related certificates and documents. Financial and legal printing is typically national, but some companies serve local and regional needs. Currency is printed by governments or by contract to financial printers.

This industry segment has experienced consolidations and liquidation, with only a few companies controlling approximately 90 percent of the volume. Automation in the form of electronic data transfers has streamlined the process of producing and distributing financial and legal documents. It is also tightly regulated by the SEC and other agencies, on matters affecting security and confidentiality.

Of utmost importance in financial and legal printing is the accuracy of information contained on documents. Security of financial documents necessitates complex technology and creative anti-counterfeiting techniques such as heat-sensitive inks, holograms, watermarks, special papers, and security threads running through currency.

Direct Mail Printing

This overlaps in many ways with other printing industry segments, notably catalogs and newspaper advertising inserts. Companies that specialize in direct mail printing often compete with publication and commercial printers that have added direct mail printing as logical extensions of their original capabilities.

Direct mail pieces are printed on a variety of paper substrates, with a marked shift toward color printing. Coated stocks can be used, but because mailing weight is a primary consideration, uncoated paper is still commonly used. Complex binding, envelope usage, and inserting are part of the workflow, with multiple printed pieces making up a single mailing.

Where direct mail has changed the most involves digital, variable data printing or VDP. Each printed piece is personalized with text and/or images of interest to a specific recipient. Marketers have found that the response rate is greater with properly-executed VDP than with non-personalized correspondence.

Digital VDP is the fastest-growing segment of the printing industry and has demonstrated the effectiveness of one-to-one marketing. The most important facet of variable data printing is setting up databases of information that allow truly meaningful personalization. This requires a different skill set than that of a designer or digital prepress technician. For sophisticated variable-data jobs, a printer must invest significant time and money in front-end technology and personalization systems.

In addition to production and data expertise, printers in this sector must be able to tune their workflow to the requirements of the postal service. Weight and size requirements have significant cost implications, as do the sorting of batches by postal code, and a working knowledge of postal distribution centers.

Direct Mail

Packaging

Package printing is a major part of the graphic arts industry. Gross annual packaging sales revenue in the U.S. is nearly equal to gross revenue in general commercial printing. Globally, packaging revenue is over $750 billion in 2017 and expected to reach $997 billion by 2020.

Packaging is the only traditional printing industry segment not adversely impacted by digital media. It includes folding carton printing, flexible packaging, label printing, corrugated board printing, and metal decorating. Digital package printing has revitalized this industry segment and has made it a focus of most digital press manufacturers.

Because packaging is such a unique facet of graphic communication, a detailed discussion of this segment will occur separately, in Chapter 10.

In-Plant Printing

This segment consists entirely of "captive" plants servicing one client, typically a parent company or institution. In-plant printing establishments range from thousands of employees to only one, but very few employ more than 100 people. In-plant printing operations include manufacturing companies with ongoing needs for print collateral, universities, insurance companies, financial institutions, and government agencies.

In-plant products range from simple business cards, collateral, and direct mail advertising pieces to complex four-color annual reports. The largest in-plant printing facility in the United States is the

United States Government Printing Office, which is responsible for printing passports, public reports, and the daily Congressional Record.

Larger in-plant facilities are equipped with some of the most sophisticated printing equipment available. Smaller facilities are typically equipped with relatively simple and small-format equipment to produce uncomplicated, routine items needed by the parent company.

Several decades ago, in-plant printing experienced significant growth through the development of small offset duplicators and presses, and simplified plate-making processes. Recently there has been rapid growth of this industry segment with the increased availability of desktop publishing and simple, print-on-demand digital presses. With the "miniaturization" and cost reduction of such equipment, companies that previously relied on the services of a commercial printer now find it economical to acquire digital equipment and produce their own printing.

Prepress Service Providers

These companies, traditionally referred to as trade shops, provide prepress services to printers not equipped to do that specialized work. Historically, these services included typesetting, mechanical art composition, scanning, photoengraving, color separation (negatives and positives), composition or stripping, and platemaking. While some printing plants elected to have their own prepress department in-house, others found it more economical to outsource.

This industry segment was changed drastically in the early 1980s with the availability of sophisticated color electronic prepress systems (CEPS) that provided efficiencies in productivity and quality that conventional prepress technology was unable to provide. One CEPS replaced the need for separate departments for each prepress step. Unfortunately, these systems were expensive. Therefore, to enjoy the benefits that CEPS offered, printers returned to purchasing services from prepress vendors that had invested in integrated systems. The service provider justified the investment by providing prepress services simultaneously to numerous printers locally or nationally.

In the late 1980s and early 1990s, the CEPS approach was supplanted by relatively low-cost desktop publishing systems. However, because many printers were reluctant to adopt the newer technology, a new type of "trade shop"—the service bureau—filled the gap. Sometimes, service bureaus were former typesetting or CEPS operations that transitioned their old business to the new technology. Sometimes, they were entirely new businesses. Most invested in high-resolution scanners, film imagesetters or platesetters, and imposition software. (See Chapter 5.) Frequently, they worked with designers to make their files ready for printing. Some offered design services for routine print projects.

Eventually, this business model changed, as printers adopted digital prepress and designers (and their software) became savvier in creating print-ready files. However, for tasks that still lie beyond a designer's capability, or are not cost-effective for a printer to bring in-house, there is still a rich opportunity for an entrepreneurial service business.

Print Brokers

Print brokers connect printing companies and customers. They are independent and not directly employed by the companies they represent; they serve as sales representatives to printing service providers and they bring in work without costing the service provider a commission. Print brokers are compensated by marking up the cost of the printing job.

Print brokers are particularly valuable to those who need printing but know little about how to specify and buy it. Print brokers coordinate all aspects of a printing job beginning with art and copy preparation and extending through prepress, press, and finishing operations. While some have basic copy preparation departments within the brokerage, brokers primarily outsource most facets of a printing job. Some brokers specialize in planning specific printed items while others are generalists and handle many types of jobs. They are typically knowledgeable in most facets of printing and know how to provide specifications. Print brokers integrate the print manufacturing process by coordinating all manufacturing services needed for the job.

5

Design and Prepress Workflow

Design and Prepress Workflow

SCAN

Chapter Preview

Pre-Digital Workflow

The DTP Disruption

Page Layout, Illustration, and Image Editing

The PDF/X-Factor

Solving Old Problems

Solving New Problems

Workflow Automation

Overview

Graphic designers and others in the business of communication have had an interesting relationship with the world of print production. In Gutenberg's time, designers were involved in the shaping of letterforms and in manually illustrating, or illuminating the printed page. Eventually, their visual images, use of color, and sizing and positioning of text was incorporated into the final, printed page—almost always by handing it off to a skilled craftsman.

As printing technology evolved, these craftsmen developed a set of skills needed to bridge the gap between the designer's vision of the page and the realities of print production and finishing or bindery. Those skills—known collectively as prepress or premedia—are just as important today as they were decades or even centuries ago. However, technological disruption has radically transformed prepress more than any other aspect of print production.

Historically, design and prepress professionals had a challenging but stable relationship. Natural disagreements and misunderstandings happened, but could

Introduction to Graphic Communication | 81

be resolved with a liberal application of good communication and professionalism. Above all, the rules of the road—what was printable and what was not—were generally understood on both sides.

That stability was disrupted by the "desktop publishing" or DTP revolution of the mid to late 1980s. Tools that were once the sole province of a skilled prepress operator suddenly became part of software that anyone could buy and use. For good reasons, designers rejoiced in their newfound abilities, but very often made mistakes, out of their ignorance of print realities. Instead of receiving artwork and instructions, prepress operators received digital files to decipher. Desktop publishing allowed the author to become the producer. Regardless of training, anyone could become a "graphic designer."

Aldus PageMaker (1985) was among the first desktop software applications to blur the lines between design and prepress.

The initial chaos of the DTP revolution has subsided. However, there is still a profound disconnect between designers using digital tools and the print professionals tasked with creating the finished product. Fortunately, automated workflows and more sophisticated applications are beginning to bridge that gap.

This chapter will address many of these issues, by explaining the underlying principles, and outlining some of the best practices needed to smooth the path from idea to results.

The Pre-Digital Workflow

Today, design and prepress functions are intertwined. Page layout, illustration, and photo editing applications contain functions from both worlds. So, to fully understand today's design-to-print process, it is important to know how artwork was prepared before the rise of digital design.

Here are the steps traditionally associated with prepress operations:

Typesetting. By the 1970s, phototypesetting systems had largely replaced the venerable Linotype machine. Instead of creating "hot type" (cast metal), phototypesetters used light beams to image letters on resin-coated or RC photographic paper. Designers or prepress professionals applied a layer of wax to the back of this output—known as "cold type"—and affixed it to...

Mechanicals. This artwork, usually on rigid board and overlaid with protective tissue, on which production notes and instructions were typically written. Type and other artwork without tonal variation (such as pen-and-ink sketches) are known as *line copy* and were affixed to the mechanical board.

A process camera held the mechanical art in a vacuum frame (bottom) and exposed a sheet of film in a separate vacuum frame (top) to create a reverse image of the artwork.

Line film negatives were made from the mechanicals, using a horizontal or vertical process camera (shown above). Line film typically contained clear windows as placeholders for...

Halftone negatives. Halftones, discussed later in the chapter, are a pattern of dots small enough to trick the eye into seeing a smooth, continuous tone image. Monochrome halftones were made with a process camera, by placing a screen over the unexposed film before shooting the continuous tone original. Color halftones were more difficult. They could be created in a process camera, but usually required a drum scanner to make...

Color separations. Typically, color halftones consisted of four pieces of film—cyan, magenta, yellow, and black—at specific screen angles.

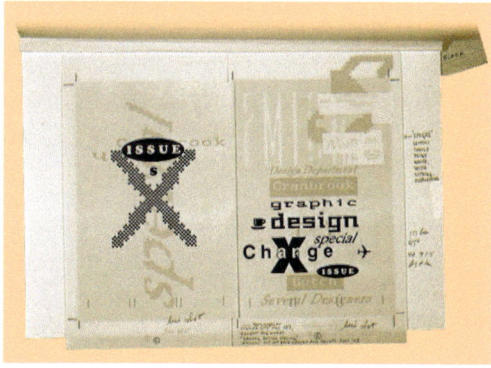

Introduction to Graphic Communication | 83

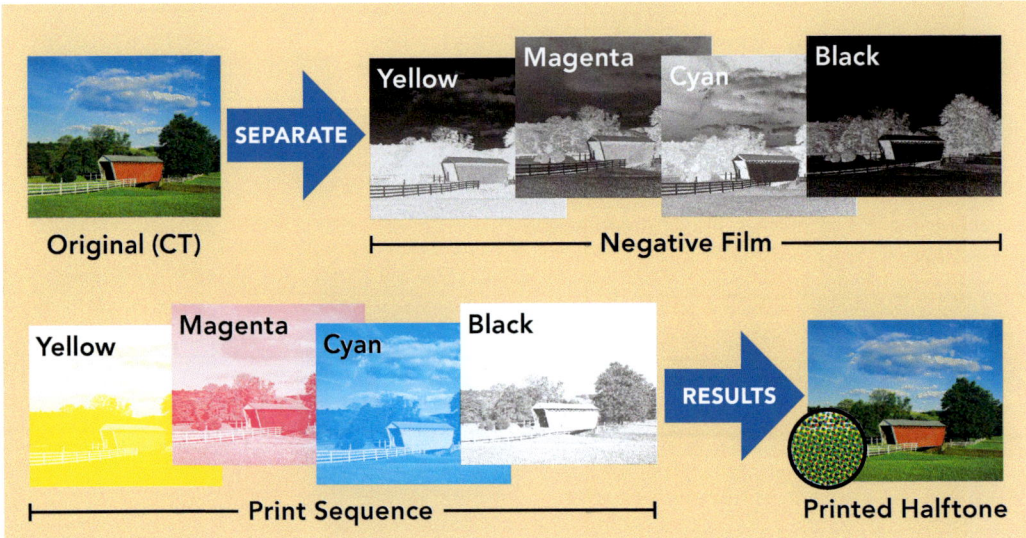

Color halftone separations, once made using analog film scanners, are now generated digitally.

When printed in sequence, these separations produce the visual appearance of a continuous-tone, color image.

(For more about color, see Chapter 6.)

Image assembly (also called "stripping" in North America) was the laborious process of mounting line art and halftone film onto opaque, typically orange paper or plastic masking sheets—punched to ensure registration and proper alignment. After taping the film to the sheet, excess portions of the mask were cut away, and red tape applied to cover flaws in the film. The result was commonly known as a flat. The opaque paper or plastic and red tape prevented light from transmitting through these areas.

Imposition was (and is) crucial to the image assembly process. Film for individual pages was expertly positioned so that the finished piece could be read after it was folded, trimmed, and bound.

Imposition is a prepress task common to both traditional and digital workflows. Correct imposition anticipates the bindery process, making sure the finished product is useable.

Platemaking was the final step in the traditional prepress workflow. A high-intensity light exposed the photosensitive offset plate underneath the flat, held in place with a vacuum frame.

84 | Design and Prepress Workflow

The Desktop Publishing Disruption

The traditional workflow imposed enormous burdens on designers and prepress alike. However, in the mid-1980s, the burgeoning personal computing industry expanded from office productivity to creative design. The new Macintosh computer popularized the more intuitive, point-and-click user interface pioneered by Xerox. It also introduced primitive (by today's standards) graphics programs with a "What You See Is What You Get" (WYSIWYG) approach to designing pages for print.

In 1985, Apple introduced a monochrome laser printer based on Adobe's page description language, PostScript (also inspired by Xerox). Around the same time, several developers introduced the first generation of page layout or "Desktop Publishing" (DTP) software for personal computers. The best-known of these was Aldus PageMaker.

Unlike its typesetting predecessors, early DTP applications did not have great typographic sophistication, and were widely disregarded by the printing industry. However, they gave page designers unprecedented on-screen control of their work—at a relatively low cost. Although designers often lacked the technical expertise to make prepress decisions, DTP programs suddenly gave them the means to do so.

The DTP phenomenon represented great cost savings potential. Many of the steps in traditional prepress existed to bridge the gap between a designer's idea and the final printed product. If a press-ready design could be created from the start, then many hours of work could simply be eliminated. For years, DTP developers sought to improve their software accordingly, adding prepress sophistication as well as designer-specific features.

> *Although designers lacked the expertise to make prepress decisions, DTP programs suddenly gave them the means to do so.*

However, design and prepress production require different ways of thinking. In many cases, designer unfamiliarity with prepress realities—coupled with ever more powerful software—has led to *increased* costs. It also led to the rise of a new "service bureau" industry *(see Chapter 4)* to cope with the problem.

Desktop design software has become much more prepress-friendly since the early DTP days. Designers and print professionals have learned how to use them to their full advantage. More importantly, design-to-print workflows

Introduction to Graphic Communication | 85

have improved significantly, realizing the benefits first envisioned 30 years ago.

The following summary of the major desktop applications will help us understand the basics of a good design and prepress workflow:

Page Layout

Aldus PageMaker had several rivals in the early days of DTP, including Ready, Set, Go!, RagTime, Ventura Publisher, and QuarkXPress. The latter overtook PageMaker as the dominant DTP software in the 1990s. However, Adobe Systems acquired Aldus in 1994, and eventually replaced PageMaker with InDesign.

By the first decade of this century, InDesign had become the dominant page layout application, with QuarkXPress a distant second. Other DTP apps still exist, including Microsoft Publisher and even Ready, Set, Go!—mainly in niche markets.

Page layout software is extremely complex, requiring much computer processing power to run well. Users import or "place" text and graphics from other programs, and subsequently position and format these elements on the virtual page. Type formatting can be controlled using global "styles" created in the program or imported with the Word file. Type behavior, such as wrapping around other objects or complex hyphenation, change in real time as other objects are moved or altered.

InDesign is now part of Adobe's Creative Cloud subscription service. It is also part of many other systems, from newspaper and magazine editorial and advertising to complex catalog production. Its core

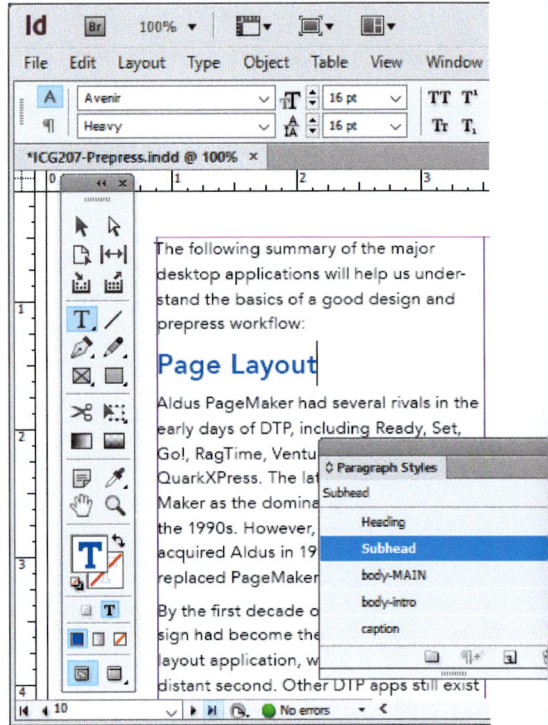

technology is also part of a separate product—Adobe InDesign Server—that is the basis for online editing systems and other Internet-enabled printing services described in Chapter 12.

InDesign also now includes features to check for potential printing errors, such as missing fonts or images. It also gives the designer the ability to export in print-optimized PDF/X format, discussed later in this chapter. Although not the same as a full prepress workflow, it does allow designers and prepress professionals to work better together.

Illustration

Other desktop software lets the user create single-page drawings. Adobe Illustrator—also part of Creative Cloud—is the dominant application, followed by Corel Draw. Less costly drawing programs are available, but may not be suited to print production.

Drawing applications typically involve **vector graphics** (from primitive geometric shapes to complex curves), as opposed to **raster graphics** (bitmap images.) The user is creating mathematical **instructions** for a shape's size, border and fill color, and other characteristics—rather than creating and altering **pixels**.

The distinction between these two ideas can be difficult. Drawing programs may let the user rasterize vector objects, and image editing programs such as Photoshop let the user create vector objects.

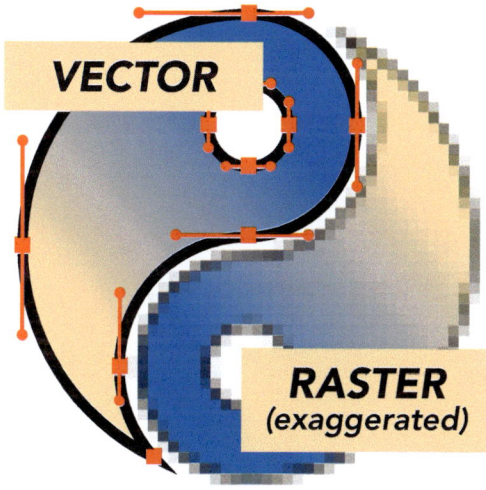

Ultimately, however, *all* graphic images—even type characters—are rasterized at the final printing stage, as described later in the chapter. Vector drawing programs simply let users to modify objects more easily before they are used in a page design.

Drawing programs, especially Adobe Illustrator, are often used in the packaging industry. This is because Illustrator's single-page approach, and its advanced object control tools, are well suited to label and package design. Packaging production systems frequently rely on Illustrator files. These system developers frequently provide specialized, Illustrator "plug-in" modules for package designers, to help manage color, and to interface with 3D rendering or CAD programs.

Image and Photo Editing

Pixel editing software is arguably the best known and most widely used application in the design world. Adobe Photoshop dominates this category, particularly when it comes to print production. Other image editing and management software tends to focus on niche markets, such as professional photography, web design, and home/hobby applications.

Photoshop is a vast, complex application, thanks in part to its widespread use in both print and digital media. Mastery of the software is well outside the scope of this book. However, when designing for print, users typically employ Photoshop to deal with:

- **Scanner input,** making sure the resulting images have sufficient resolution, are in a print-friendly file format, and are not overly compressed.
- **Digital camera files** (same issues)
- **Color adjustment,** optimizing images for printability—including contrast, tonal values, and other aesthetic decisions.

Photoshop does offers image-specific color management capabilities, such as profile-based conversion of RGB to CMYK. However, designers and print professionals should rely more on high-end color management tools, as discussed in Chapter 6.

The PDF/X-Factor

There are many other software tools used widely across the design and prepress worlds, including Microsoft Office (mainly Word) and many traditional and cloud-based applications and utilities. However, the single most significant digital technology affecting design-to-print workflows is the Portable Document Format or PDF.

PDF was originally developed by Adobe as a universal document format—readable on any computer, even if the original authoring application was not installed. A PDF file retained all the appearance of the original, and could be shared, opened, and printed anywhere, using Adobe's free Reader program.

Since its introduction in 1993, PDF has undergone many changes, and has been appropriated by many, sometimes competing interest groups. Adobe controlled the specification until 2008, when it was released as an open ISO standard.

Because it is a self-contained digital description of a page, PDF was widely embraced by the printing industry. Film and plate output vendors—already using Adobe's PostScript language—began to consider PDF as a more efficient interim format. But where it made the greatest impact was in the digital workflow from designers and agencies to their print production counterparts.

Early users of PageMaker, QuarkXPress, and InDesign often submitted "native" application files to printing companies—with disastrous results. Issues such as missing fonts and low-resolution images (discussed later in the chapter) put a heavy burden on service providers. Using PDF, a self-contained file format, was the potential solution.

Early versions of PDF did not fully support the needs of prepress professionals. However, efforts by leading prepress developers led to the formation of the Ghent Workgroup (GWG) and ultimately a standardized version of the format for prepress file exchange: PDF/X.

Today, PDF/X files contain all the necessary production elements of a print job, including the right fonts and images, page geometry information such as trim and bleed, and even color space information discussed in Chapter 6.

The current version of the ISO standard, PDF/X-4, facilitates the idea of "blind" file exchange. In other words, prepress technicians—and the applications they regularly use—are not faced with unknown output factors affecting the printability of the job.

The format is not yet finalized. (PDF/X-6 is under discussion at the ISO.) However, since InDesign can export jobs in PDF/X, the format is now the basic workflow "building block" for communicating a designer's intent.

Solving Old Problems

Many traditional prepress tasks are still important in a digital workflow. This is because printed jobs are still a physical object—usually paper. Ink and toner properties also can create results that are not immediately apparent on screen. Also, the finished print product still has to be cut, folded, and bound for optimum consumer use.

PDF/X files and workflow systems can sometimes avert or at least detect issues related to ink, toner, or finishing. However, designers are still very much in control, and their decisions may adversely affect the printed outcome.

Ink behavior is one of the major sources of miscommunication and error. The issues include:

INK COVERAGE

C = 100%
M = 100%
Y = 100%
K = 100%

400% Total Ink Coverage (TIC) creates *many* printing problems.

OVERPRINTING

Magenta O/P Cyan

Magenta O/P Cyan

Overprinted objects may look different on screen than in print.

Although ink percentages and overprinting can be controlled by the designer, the printed results may be disappointing. (Note: To avoid printing problems, the <u>actual</u> CMYK values for the "rich black" shown here are C=60, M=40, Y=40, K=100, for a TIC of 240%.)

Total Ink Coverage or TIC. Print color is often a combination of cyan, magenta, yellow, and black inks, or CMYK. For darker colors, such as a "rich black" background, designers may be tempted to use 100 percent of all four colors. This results in a TIC value of 400 percent.

Such a color can look fine on screen. On the printed page, however, it creates printing and drying problems, such as offsetting the too-rich-black onto the back of each sheet in the stack. It also wastes ink without really improving the visual appearance.

TIC limits for photographs are usually set in color management systems discussed later in the book. For manually-created tints, TIC should not exceed 300 percent.

Overprinting. To avoid registration problems, designers may set one colored object to overprint another, rather than "knock out" or remove the background color. This is fine for black or other dark objects on a lighter background. However, unless the desktop application is set to preview the overprint effect on screen, designers may be seeing something quite different from the final print.

Bindery issues (covered in Chapter 9) also have impact on print design. They include:

Trim and Bleed. Print jobs are nearly always trimmed to their final size from a larger printed sheet. However, some sheets shift during the trimming process, so the results are seldom 100 percent accurate. To compensate, designers must always add a bleed area for background elements.

The *trim size* of a printed piece (black line and crop marks) are chosen by the designer. Page **bleed** (red line) and a safe area for key elements (white line) compensate for shift during printing and trimming.

The page bleed margin is represented as a red outline in InDesign, and will be part of the "bleedbox" definition in a PDF/X file. However, it is always up to the designer to make sure that background tints or images extend past the trim and at least to the bleed margin.

Trimming inaccuracies are not the only reason to add bleed. Some printing devices do not hold tight registration, so a slight shift may occur from page to page. (For this reason, designers should also avoid placing critical elements too near the trim edge.)

Imposition. Digital prepress has transformed page imposition into a highly automated process. If designers export press-ready files (typically PDF/X), then software programs such as Impostrip or Preps literally do the math and position each page, according to the press sheet size and binding chosen for the job.

For the imposition of books and multi-page publications, the designer's job is relatively simple. Bleed objects must be correctly placed, and single-page, PDF/X output (not reader spreads) must be selected during file export.

Adobe InDesign and QuarkXPress both provide PDF/X-4 export capabilities. However, not all desktop applications are as prepress-friendly. When using other design programs, even simple imposition can be a challenge.

For simpler jobs, such as brochures or greeting cards, imposition may create challenges for the designer. High-end imposition software is not typically used, so designers may need to create a hand numbered "folding dummy" (see below), and use it to plan page order and rotation in their design application.

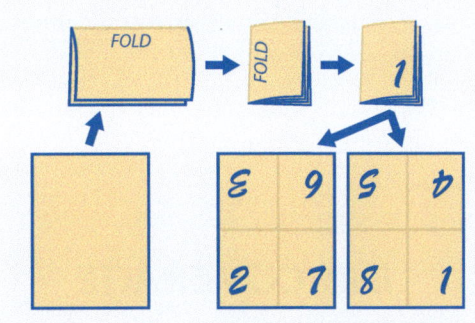

Although digital technology has taken most of the manual labor out of the imposition process, designers must still remember that the order of pages on their screen is not how they must be positioned in a printing and binding environment.

Halftone Screening and Tints. Since the idea was patented in 1852, printers have been using patterns of dots to simulate a continuous-tone image. Digital technology has not eliminated the halftone; it has made the process more complex.

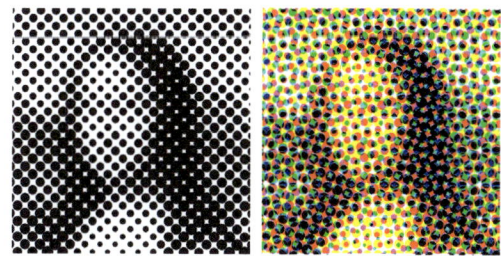

A halftone is the breaking down of gray or process color values in a continuous-tone image into solid dots of various sizes. Smaller dots represent the highlight or lighter areas, middle-sized dots represent the mid-range tones, and larger dots represent the shadow areas or dark tones.

Though different in size, the dots are equidistant from center to center. In other words, the frequency or number of dots per linear inch is the same. When viewed, an optical illusion occurs. The viewer cannot see the individual solid dots because the eye blends the solid dots with the white paper around the dots, and the image resembles the shades of the original, continuous-tone image.

(In the exaggerated example above, the optical illusion only starts to work if the page is viewed from a greater distance.)

For offset reproduction, halftone screening required that each process color dot pattern be set to a specific angle. More than one combination of screen angles is possible, but the goal was always to produce a pleasing, non-distracting pattern. Under magnification, a correctly printed color halftone would have a regular "rosette" pattern of colored dots, but would resemble a continuous-tone image to the naked eye. If incorrect screen angles were used, the image would appear to vibrate or *moiré*—an unsettling visual distraction.

Historically, halftones were produced photomechanically, using large graphic arts cameras or drum scanners. Now, they are almost all generated digitally, translating color or grayscale values in a digital file to precise patterns of extremely small dots, called rasters.

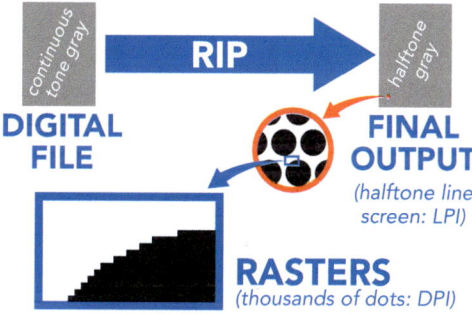

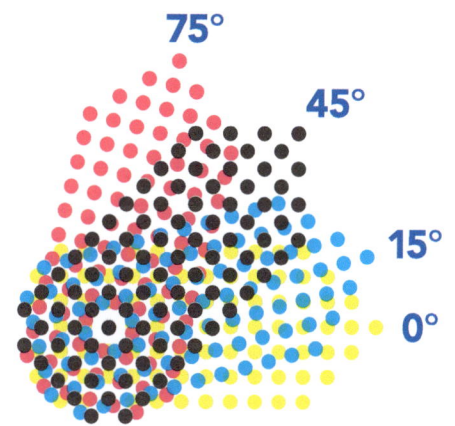

The system for doing this is known as a Raster Image Processor or RIP. These were originally developed to process PostScript files, but now handle PDF. A RIP is standard to all digital film imagesetters, platesetters, and digital presses, and are used to control a laser imaging, inkjet, or other output device.

The raster marks created by this system are very small—typically 2,400 dots-per-inch (DPI) or more. These are grouped to create type, line art, and halftone images, using the traditional designation of lines-per-inch (LPI).

Traditional halftone dots are possible but certainly not required for digital output. Most inkjet and toner-based devices use proprietary screening to produce nearly continuous-tone output.

A common, non-traditional halftone screening method—often used in digital printing, but also possible with conventional printing—is known as **stochastic** or Frequency Modulated (FM) screening.

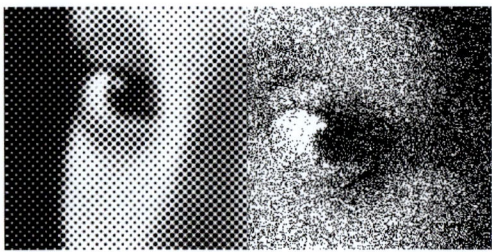

Halftone screening types, exaggerated: conventional or AM (left) and stochastic or FM (right).

With conventional screening, the halftone dots occupy fixed positions, but vary by size—large for dark areas and small for light areas of the image. Stochastic screen dots are all the same size (i.e., very small) but there are more of them in dark areas and fewer in light areas.

The terms "AM" (Amplitude Modulated) and "FM" mean that one represents light and dark areas by dot size or **amplitude**, and the other by dot **frequency** and position on the page.

FM screening produces print image quality comparable to continuous-tone photography, without *moiré* and other conventional halftone hazards. However, when used in conventional offset or gravure printing, it requires tightly-controlled press conditions and the ability to consistently print a very small (25 micron) dot—half the width of a human hair.

A variation of the halftone is the screen tint, typically used for backgrounds, and often requiring a bleed margin. Tints consist of dots of a particular ink color at a particular size or frequency.

In the early days of desktop publishing, gradient fills (the smooth transition from one color tint to another within the same object) became very popular. Previously, these could only be achieved using manual airbrush techniques, and were both expensive and time-consuming. Desktop design applications made them far easier to create, although early RIPs often had difficulty rendering them. Modern RIP technology has fewer problems with gradient fills, which has led to their widespread use (or perhaps overuse) in graphic design.

Tints allows the creation of many different color values and many different effects from a small number of process colors. As with halftones, tints are subject to color shift under different press conditions and therefore require color management, which is covered in Chapter 6.

For designers and prepress technicians, the options for halftones and tints have grown exponentially with the rise of digital technology. Thankfully, computers handle the complex math needed to generate the dots. But it is still up to the designer to use photographs and tints properly under multiple print conditions.

Solving New Problems

Digital design tools gave designers unprecedented control over production issues that were formerly the province of typesetters, color separators, and other prepress professionals. Since the early days of DTP, this has been a source of problems. Designers' lack of prepress knowledge, coupled with early-stage software deficiencies, resulted in unprintable files and disappointing results.

The short-term solution was the service bureau industry. Typesetting companies and color separation vendors, whose business was being undermined by DTP technology, shifted to solving problems created by early DTP workflows. However, the emphasis soon turned to software solutions, notably standalone "preflight" applications and eventually robust error checking within DTP apps themselves.

The cost of fixing file problems varies greatly. During design, it takes only minutes, but if a problem is discovered on press or after a job is printed, the time and cost impact can be catastrophic.

Preflight applications, such as Flight-Check, PitStop, and pdfToolbox analyze PDF files, detecting (and in some cases fixing) conditions that would result in failure. FlightCheck also works with "native" QuarkXPress and InDesign files.

> *The preflight process assumes that designers and print service providers are openly communicating.*

The preflight process assumes that the designer and print service provider are openly communicating about practical, print-related issues. This can take several forms, from ordinary conversations to detailed lists of requirements.

The list of typical file problems is largely the same as it was years ago. Here are the most common issues:

Missing or improper fonts. Designers have virtually unlimited choices when it comes to typeface selection. However, if the printer does not have the same font as the designer, it may be replaced with a generic font—or may fail to output at all.

With PDF/X output, fonts are automatically included. However, if a program does not support PDF/X, then the designer must supply all fonts used, or convert text to outlines or paths before sending the file.

Wrong color mode. Within a robust color management ecosystem (see Chapter 6), RGB images are permissible, or even desirable. However, without such a system, designers who

Visual concept courtesy of Markzware

submit RGB content to a CMYK printing environment will often be disappointed.

Low effective image resolution. The pixel dimensions of an image affect how well or how poorly they will be reproduced. Below a certain threshold, typically about 150 pixels per square inch (PPI), a printed image may appear blurred or pixelated, because the RIP simply did not have enough data to create a sharp, clear image.

The problem is made worse when an image is enlarged. If a 300 PPI image is stretched from two to ten inches wide, then its effective resolution is only one-fifth of the original, or 60 PPI—too low to produce a good image.

Missing or unlinked graphics. InDesign and QuarkXPress keep track of placed or imported graphics, but do not embed them in the layout file. If the link to these external files becomes invalid (by moving or renaming a file, for example), then the RIP will not be able to process and output the page.

The OPI Workaround

Before the advent of higher-speed networks and less expensive data storage devices, large image files posed a serious problem for digital prepress. The solution was to create low-resolution "proxy" or placeholder images, use them in the layout application, and replace them during the final output stage.

This workflow, developed by Aldus, was called Open Prepress Interface or OPI. As design and prepress systems became more robust, OPI has largely been abandoned.

Finding and re-linking missing graphics is standard procedure for InDesign and QuarkXPress—as is "packaging" related files when sending the layout to a printer. Exporting a PDF/X file also fixes the problem, although the resulting files can be extremely large.

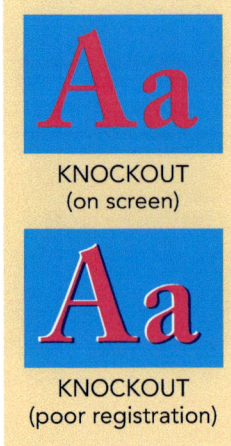

Elements set to overprint (left) or knock out (right) may seem correct on screen, but will combine colors in one case and risk creating white gaps in the other, if press registration is poor. *(Misregistration is exaggerated here.)*

Overprint and knockout. Colored elements on a page can be set to either overprint or remove (knock out) the color of the element behind it. Overprinting can inadvertently change the color results, as discussed on page 89. Knockout creates another potential problem. If registration is not perfect, there may be a white gap between color elements. This can be remedied by adding an overlapping "trap" color to compensate for the shift. (This process is not the same as ink trap, which is described in Chapter 8.)

Multiple spot colors. Desktop applications allow designers to specify unlimited spot colors. However, the number of actual colors available on a press is limited. Besides the four process or CMYK inks, presses can only affordably use a handful of custom or spot colors at a time. Designers must therefore convert extraneous spot colors to process, or allow their printer to do so. Again, this should always be done in the context of a good color management environment.

Bleed area and page size issues. As noted on page 90, printing and paper trimming can be imprecise. Unless the designer extends background elements the required amount beyond the page edge, there is a risk of a gap on one or two sides of the page. This area, known as the page bleed, is identifiable in many desktop applications, but the designer is ultimately responsible for making sure bleed areas are filled.

A seemingly obvious issue is that of page size. Printers and designers should always have a clear understanding of the ideal page dimensions for a particular project, but the designer is typically the one who specifies this at the outset.

Page layout applications have added sophisticated features for detecting many of the issues described in this chapter. Dedicated preflight applications go even further, making it less likely for designers to inadvertently produce files that will cause a job to print incorrectly. Increasingly, error checking is becoming part of internet-enabled workflows, as discussed in Chapter 12. However, the best defense against catastrophic print job failures are designers who understand print.

Workflow Automation

The DTP phenomenon broke down the traditional work distinctions between designers and prepress technicians. Although the software gave rise to problems, it also opened up new possibilities for automation. The digital file—specifically PDF/X—became the common denominator for a unified, assembly line-style approach to creating and producing printed materials.

Unlike other mass-produced products, print retains a very high degree of customization. There are many variables, beginning with page size, paper types, press conditions, and binding styles. Print designers and producers also benefit from creating products that are distinctive and unique, and are always seeking new ways to stand out from the crowd.

As a result, the rules for print automation must be extremely complex—or else limited to more predictable types of printed jobs.

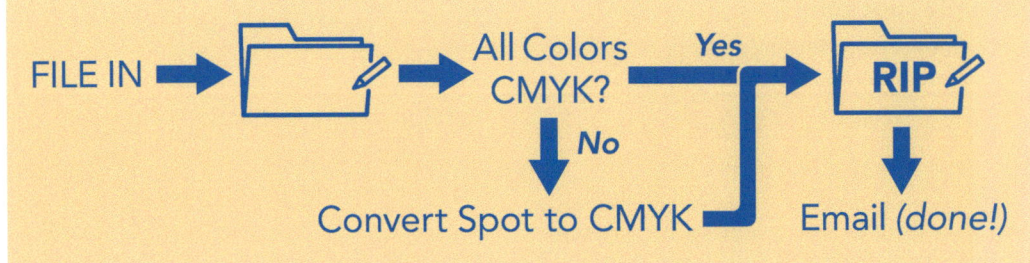

In the late 1990s, the Adobe prepress group proposed a workflow automation system based on prepress applications' ability to execute routine tasks without user intervention. Called Adobe OPEN, it allowed users to drag files to network "hot folders," triggering events within a prepress application, and then passing the file along to another program.

In theory, this would automate many routine prepress workflows, reducing or eliminating repetitive labor costs.

Automation Engine (Courtesy of Esko Graphics)

OPEN was unsuccessful, but it marked a trend towards workflow systems that resemble pipelines or transit systems. Each module or junction represented a task to be accomplished, based on a series of yes/no or true/false conditions.

Today, examples of this approach include Apogee, Automation Engine, Prinergy, PuzzleFlow, and Switch. Larger printing companies have also developed their own workflow management and automation systems, such as RR Donnelley's InSite system.

Very often, these workflow systems are coupled with electronic job tickets, allowing print buyers to specify details of a job in a way that humans and digital systems can comprehend. The best known electronic job ticket approach is embodied in the CIP4 Job Definition Format (JDF), covered in Chapter 11.

Another important facet of prepress automation is the rise of internet-enabled print, which is the subject of Chapter 12. The business of design and print is increasingly conducted online. So, the tools for interconnecting the complex process of graphic communication will continue to evolve—on browser-based systems and mobile devices connected to secure web services.

However, automation will never eliminate the need for creative design or skilled execution in print. Repetitive tasks—and the manual labor they require—will be supplanted. Creativity, skill, and innovation will remain.

6
Color Management and Proofing

Color Management and Proofing

Chapter Preview

Why Manage Color?

Adding and Subtracting

Basic Color Models

Extended-gamut and Spot Color Inks

Gray Balance and G7

Color Proofing Overview

Inkjet Proofing

Soft Proofing

Overview

All graphic communication relies on the visible spectrum of light—that narrow band of electromagnetic radiation that can be seen by the human eye. Whether our communication is based on text, images, or moving objects, the colors we choose for each element affect how our message is perceived.

Color is both objective and subjective. It can be measured with extreme accuracy, but it also has psychological and cultural effects. For example, some maintain that the color red connotes energy or courage, while green signifies growth or renewal. Colors also have widely different meanings to people of different cultures.

The subjective and cultural aspects of color are beyond the scope of this book. Instead, this chapter will focus on the scientific principles of color, and on the practical aspects of color management. It will also include the basics of proofing—the process of reaching an agreement on the color of a printed piece.

Why Manage Color?

For an artist creating a single, unique piece, color is a matter of selecting a particular canvas or other surface, and choosing the desired pigments, paints, pencils, or other substances used to express an idea.

For a one-to-one communication such as a painting or drawing, color choice is straightforward—and entirely up to the designer. However, to communicate with a large audience, over multiple print channels, there must be an *economical* way to reproduce color with accuracy and consistency.

For example, an artist may choose a particular shade of umber for a piece. However, to create a printed or online replica of that color, it would not be economical to create that exact shade of umber ink—or invent a way to generate an umber-colored pixel.

Instead, the solution is to *simulate* umber with a combination of primary colors. In the world of print, that means using the three *subtractive* colors: cyan, magenta, and yellow (CMY) with the addition of a key color or black (K). For broadcast and online media, it means combining the three *additive* colors: red, green, and blue (RGB).

Color management is all about making those combinations with predictable and repeatable accuracy—honoring the designer's original intent no matter what output device is used to mass-produce the work.

A Matter of Balance

Before delving into color models, it's important to understand a key concept: *gray balance*. In photography, a neutral gray (not too warm, not too cool) is achieved by combining certain percentages—not necessarily equal percentages—of RGB. Once that state is achieved, through lighting and camera adjustments, then all other colors can be accurately captured.

The same holds true for print. While CMYK is the basic model for print, keeping color consistent has become increasingly difficult. The number of print output options is expanding rapidly, making it harder to reproduce a designer's original color intent on every device. However, if a measurable gray balance can be achieved, then traditional color management techniques (pages 128-129) do not need to do as much "heavy lifting" to get the color just right.

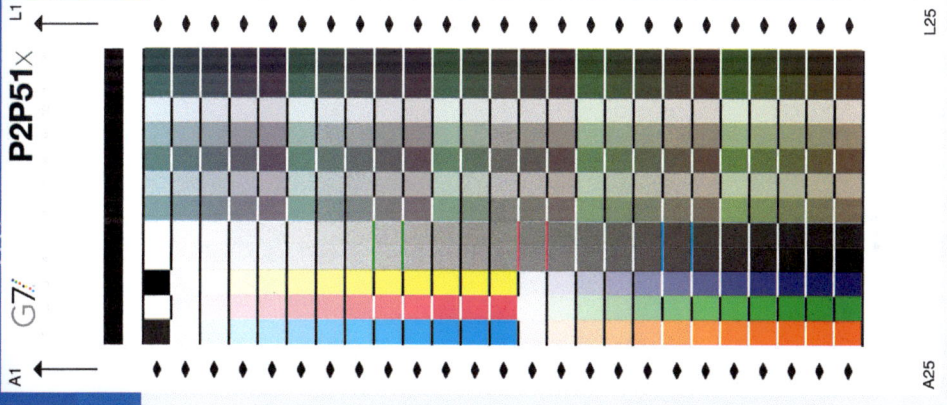

Measuring a printed "P2P" target is part of the G7 process for achieving consistent gray scale appearance. (Image courtesy of Don Hutcheson, www.hutchcolor.com .)

The way to achieve gray balance in print is the G7 process described on pages 130-131. For now, think of it as a way to achieve a neutral gray (not too warm, not too cool) so that other colors can be printed accurately with *any* type of printing (digital, offset, flexography, gravure, screen, wide format, etc.) to maintain design and brand consistency.

Color Basics

There are millions of color variations visible to the human eye and interpreted by the brain. However, to reproduce these economically in print requires a shortcut—combining a small number of basic colors to simulate as many other colors as possible. All of color printing is based on this principle, combining (typically) cyan, magenta, and yellow inks—with the addition of a key color, black. This is commonly abbreviated as CMYK, although those four colors are not always printed in that order.

The use of CMYK, also known as process color printing, is required to reproduce color photographs, illustrations, and solids or tints of any other color. Black ink—typically less expensive than the other process colors—is used to supplement CMY, and to improve image quality and detail, as well as printing type and other black-only elements of a page.

> **The theory of process color is simple. Applying it consistently, over multiple devices and conditions, can be difficult.**

As discussed in Chapter 5, photographic images and color tints are rendered as a pattern of dots, known as a halftone. The purpose of the halftone model is to economically reproduce color in a way that appears to be a continuous-tone background or photograph, by printing each color separately and in sequence. However, while the theory of process color separation and printing is simple, applying this theory consistently, over multiple devices and conditions, can be extremely difficult.

USED IN THIS BOOK:
Cyan = 85% / Magenta = 50%
BACKGROUND:
Magenta = 15% / Yellow = 30%
(TYPE: Black = 100%)

Adding and Subtracting

When viewing a color on a monitor, the eye is receiving a combination of red, green, and blue light wavelengths, each emitted by a very small crystal or diode that has been electronically activated. (For older computer or television monitors, the color principle is the same, but the color source is from red, green, and blue phosphor dots on the surface of a glass tube.) The appearance of color is created by *adding* specific levels of the primary colors.

Printed color relies on *subtracting*, rather than adding color frequencies. Once applied to paper or other substrates, process color inks are so thin that they are transparent. They act as filters that transmit reflected light, reducing certain frequencies and letting others get through. By overprinting yellow, magenta, and cyan transparent halftone dots at prescribed angles, nearly all colors of the image or background can be replicated. For example, yellow and magenta overlap to produce red; magenta and cyan produce blue; and yellow and cyan produce green.

Another way to view subtractive color is to start with the paper, which is typically (but not always or even purely) white. By definition, a white surface reflects most of the visible light spectrum it receives, namely equal amounts of red, green, and blue. **Process colors absorb their opposite light frequencies and reflect the others.** For example, solid yellow ink absorbs its opposite (blue) and allows equal parts of red and green to be reflected from the white paper surface.

BASIC COLOR MODELS

In the RGB color model, specific levels (from zero to 255) of each primary color are added—and transmitted to the eye—to simulate millions of visible colors.

The CMY model uses percentages (zero to 100%) of primary color-pigmented inks. When combined, these inks subtract certain color wavelengths of the light reflected from the paper or other substrate.

The Color Challenge

The difficulty of predicting and controlling millions of colors is obvious, especially in light of our highly subjective interpretation of color and its meaning.

Color management is also a moving target. What was once acceptable in a print or online medium years ago may be intolerable today. It also depends on the graphic communication venue. Color requirements for advertising or product packaging are different than those of a newsroom or office. Good quality is always desirable, but the need for precision varies.

Color management must deal with several inherent challenges, including:

Human perception. Not everyone sees color the same way. For example, studies have shown that men are less adept than women at distinguishing among shades in the center of the color spectrum: blues, greens, and yellows.

Lighting conditions. A difference in ambient light sources (indoor vs. outdoor, morning vs. afternoon, incandescent vs. fluorescent, etc.) will change the perception of color. Graphics professionals typically use a D50 light booth as the ideal viewing environment.

Substrate differences. Some papers include optical brighteners and other whitening agents, while others do not. Some printing surfaces are not white at all. Under the subtractive color model used in print, the composition of paper and other substrates has a direct effect on the color result.

Despite these difficulties, color management has progressed rapidly. Graphic communicators can now design and produce printed and online color with greater confidence in its quality and consistency than in the past.

Advanced Color Models

The mathematics of color theory are not covered in this book. However, it is useful to know some of the basics, which are often used in design applications, such as Photoshop, and in proofing systems.

The International Commission on Illumination (CIE, for its French name) was established in 1913 as an international, nongovernmental authority on the science of light, illumination, and color.

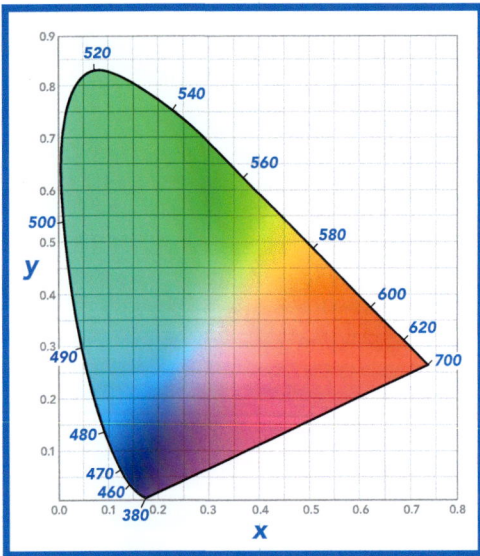

The CIE 1931 color chromaticity diagram. *(Color appearance in this printed diagram is different from the actual colors.)*

The CIE 1931 color space model defined color mathematically, based on the wavelengths of visible colors and their perception by the human eye.

The familiar, arch-like diagram arranges color by wavelength—from 380 to 780 nanometers (violet through red). It can also illustrate the range of color that a particular device or process is capable of displaying. This color range or **gamut** can vary widely from device to device.

The three-dimensional CIEL*a*b* color space.

The CIE also developed another color model for describing color—and the respective color gamuts of input and output devices. Known as CIEL*a*b* (or CIELAB), it defines color on three axes: "L" for lightness, "a" for the red-to-green color range, and "b" for the blue-to-yellow color range. (L*a*b* measurements are typically those captured by a color spectrophotometer.)

In order to reconcile the differences between color output devices, the International Color Consortium (ICC) established standards for color device characterization.

An ICC profile is a set of data that defines a device's color attributes and makes it possible to "map" colors between different devices. This works well for users of Adobe Creative Suite and other

ICC profile-aware applications. Correct profile use can minimize the color differences between software applications, and on different RGB displays.

ICC profiles are also used for print workflows, but with important caveats. The color gamut of process color printing is often smaller than even a basic color monitor, and usualy much smaller than that of a high-quality display.

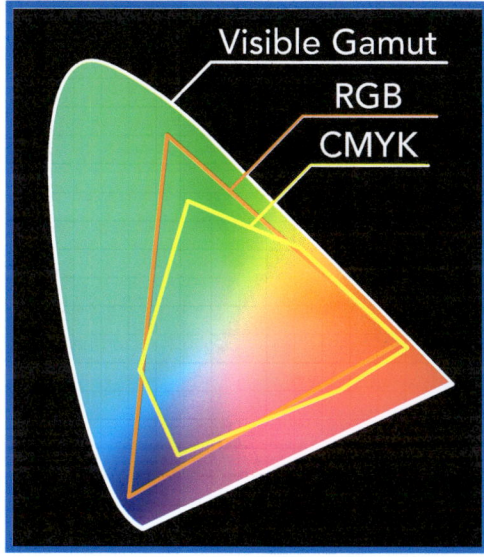

The color gamut of CMYK printing is often smaller than that of an RGB device, making certain colors difficult to reproduce.

An ICC workflow can "map" a particular color in RGB to its nearest CMY value. However, with colors that are outside the print color gamut, the results can be unsatisfactory. Using an ICC profile approach to mapping colors on multiple devices can also create other problems. With multiple devices and profile connections or links, it poses a significant workflow management problem, and a risk of color conversion error.

Extended-Gamut and Spot Color Inks

One common solution to the problem of printing out-of-gamut colors is simply to use different inks. Some print systems—notably wide-format inkjet devices and high-end digital inkjet presses, add special orange, green, or violet inks to the usual CMYK. This approach widens the process color gamut of the device.

Extended-gamut color (sometimes known as expanded-gamut) is used for a variety of print applications, especially product packaging, where color is a major factor in catching a shopper's attention, and where printing conditions require limitations on total ink usage. Offset, flexographic, and digital printing presses (see Chapter 8) often employ additional color units for just this reason.

On many occasions, a particular color may not be easily achievable using CMYK, or even with the use of extended-gamut inks. These can include brand-specific colors (e.g., "Coca-Cola Red") or other design colors, such as metallics and fluorescents.

When such a color is needed, the solution is to use a custom or spot color ink. Its chemical makeup is literally the hue needed by the designer or brand owner. For example, a spot ink with a specific shade of orange might include a mixture of red, yellow, black, and white pigments.

These special-purpose inks are manufactured by companies such as Pantone, giving rise to the common association of the phrases "spot color" with "Pantone color" and "PMS color." (The latter is an abbreviation for the original Pantone Matching System.) In addition to its

Pantone-manufactured spot colors have a specific formulation, combining base inks to create over 1,800 separate colors for coated and uncoated stock. The company also publishes a library of process colors, approximating spot or custom colors in CMYK or extended-gamut combinations.

library of spot color inks, Pantone also publishes their approximate CMYK, RGB, and HTML values in a separate guide, although these numbers do not guarantee accuracy in an environment with multiple output devices.

Some companies have very specific ink formulations for their brand colors. These colors can be trademarked, to discourage competitors from copying their brand appearance, but not patented.

Spot or custom colors are not foolproof. Different ink formulations and press conditions can create unacceptable deviations. So, in 2000, GretagMacbeth (since acquired by X-Rite) developed the Color Exchange Format (CxF) specification—a way to define and communicate color *scientifically*, using spectral measurement values. The ISO adopted CxF as a standard in 2015.

The CxF workflow, as with other approaches to color management, relies on the measurement of a color's spectral, L*a*b* values. Once the values for the solid and tint percentages of a spot color are measured, the CxF profile data can be included inside a PDF/X file (to control output behavior) or used to formulate the ink correctly. In some cases, CxF-based color palettes are used in Adobe Illustrator to ensure design consistency.

Shades of Gray

In process color printing, black ink serves a specific purpose. Merely combining equal parts of cyan, magenta, and yellow typically does not produce a satisfactory black, and is prone to undesirable color shifts in gray areas of the page if the CMY combination is not exactly right.

CMY-combination black is also more expensive, based on ink cost. Therefore, to reduce costs and improve image quality—particularly in the darker or shadow areas—four-color separations often replace equal parts of CMY with a calculated amount of black.

This process, known as gray component replacement, or GCR, has been common industry practice for years. What is less common, until recently, is the use of gray balance in color management generally.

In 2006, the industry association Idealliance updated the ICC profiles for web offset printing of publications (SWOP) and commercial sheetfed offset printing (GRACoL). At the same time, it established a new calibration method, called G7, to ensure a common neutral appearance, regardless of inks, substrate, or printing technology.

G7 is not a color management system, *per se*; it's a gray management system. By using simple CMYK tone curves in the RIP of a platesetter or digital press, the process allows any device to produce consistent, truly neutral gray output. Once a device is so calibrated, then profile-based color management has a consistent foundation for keeping all devices color-accurate and consistent.

 Many digital presses and other output devices also have their own, built-in calibration, designed to bring it back to original "factory specs." This can serve as an additional foundation for G7.

Idealliance's G7 Master Qualification is a three-tiered program for printing companies—renewable annually—and individual color professionals. G7 Grayscale (fundamental), G7 Targeted (intermediate), and G7 Colorspace (advanced) each show a different level of competency,

COLOR FOUNDATIONS

By establishing a common neutral appearance, G7 methodology facilitates more efficient profile-based color management. (Illustration courtesy of Idealliance.)

capability, and color management aptitude, as well as help print buyers know whether a printer has achieved the appropriate G7 level for a particular job or a particular process.

Idealliance also offers a G7 System Certification Program, which evaluates a system's ability to calibrate a printing device to meet G7's grayscale definition.

Note that the G7 Master label does not guarantee that the process has been used in a particular job, but rather that a plant or individual is qualified to do so, using G7 Certified software.

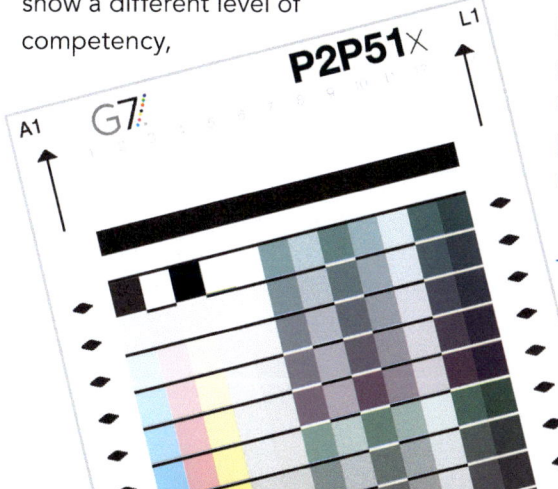

In the G7 process, a P2P color target is printed out and measured, using a spectrophotometer. (Image courtesy of Don Hutcheson.)

Color Proofing and Prototyping

In Chapter 5, we reviewed the many tasks that designers and production professionals must perform correctly in order to produce a successful print outcome. One of these, which depends on a basic understanding of color, is the process of proofing, reviewing, and approving a print job—*before* it goes on the press.

Not all jobs require proofing. With the advent of desktop design and print-on-demand, a large percentage of work is simply approved on screen, by sharing a PDF or other digital file with designer or print buyer. However, if matching color is the main concern, then some form of proofing must be employed.

A Brief History

Before the rise of digital prepress, color proofing was done on dedicated proof presses—scaled-down versions of actual printing presses—at a significant cost to the publisher or advertiser. Typically, proof presses were limited in quality.

In most cases, it was more feasible to send a representative to the printing plant, to approve the job as it was being set up on press. This process—called a **press check**—required much time and travel costs, as well as creating scheduling constraints. Very often, the cost of doing press checks could only be tolerated for high-end work.

Early color proofs could be created using color overlays (positive CMYK film layers taped over white artboard). But although they cost less than a press check, overlay proofs did not always simulate color to a designer's satisfaction. They did not qualify as a **contract proof**—a signed, legally-binding sheet indicating the designer's or art director's approval to print the job.

A major breakthrough in modern color proofing was the laminate proof, such as 3M's Match Print and Dupont's Chromalin systems. For the first time, a single sheet could be produced to represent the finished, printed product.

Changes in digital technology also spurred advancements in color proofing, including dye sublimation printers like 3M's Rainbow and Kodak's DCP systems in the early 1990s. Later in that decade, laser sublimation systems like Kodak Approval and Fujifilm FinalProof created contract-quality color proofs—with halftone screening—but at a cost that many considered prohibitive.

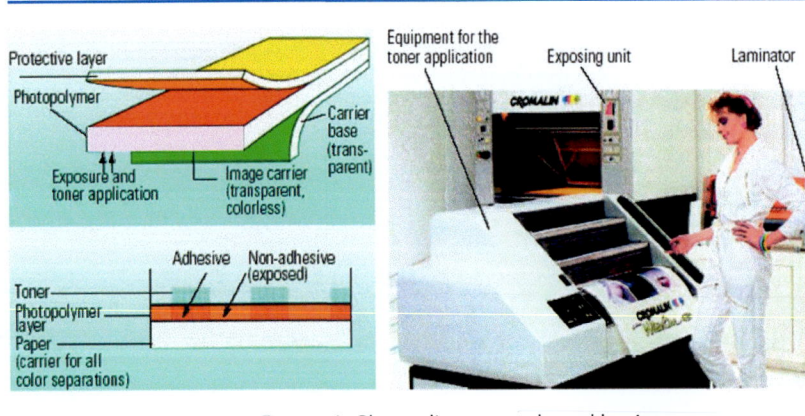

Dupont's Chromalin, a toner-based laminate proofing system, was developed in the 1970s.

The Inkjet Revolution

The most significant advance in digital color contract proofing involved inkjet printing technology—described in more detail in Chapter 8. Inkjet's ability to produce a single-color page at the same cost-per-page as a longer run made it a logical choice as a proofing technology. But the use of inkjet for proofing was problematic at first.

An early inkjet system used for proofing was the Scitex Iris printer, which used continuous drop technology, not drop-on-demand. Although capable of brilliant color output for art prints and similar applications, its high cost and maintenance requirements limited its widespread adoption as a proofing device.

The lower-cost alternative, drop-on-demand inkjet, could also produce vibrant color—on a much larger sheet than most electrophotographic, toner-based devices. However, inkjet's vivid colors, well suited to posters and signage (their original application), could not easily be controlled for contract proofing.

In the early 2000s, several developers began selling color-managed RIP and workflow software that could make the inkjet device precisely emulate color output on an offset or other conventional press. These vendors included Agfa, Bestcolor (now owned by EFI), and CGS-ORIS. Such systems were later expanded beyond proofing to control final color output on digital presses.

Today, making contract-quality proofs on an inkjet device can be as simple as clicking "Print," or exporting a press-ready PDF/X file. The actual inkjet device may be at the printing facility or installed

Accurate contract proofs can be printed on color-managed inkjet devices. (Images courtesy of Epson.)

at a publisher or other customer site. In the latter case, known as remote proofing, the prepress or premedia operator still makes the color decisions, and the output is generated via PDF/X files sent to the remote system.

> *Making contract-quality proofs on an inkjet device can be as simple as clicking "Print."*

Inkjet proofing has progressed considerably since its inception. Because inkjet devices must print on specially coated sheets, proofing vendors have created a variety of new substrates for color-managed proof output. These simulate a

Monitor-based proofing requires a display that can be physically calibrated and shielded from environmental lighting conditions. (Image courtesy of Eizo.)

broad range of coated and uncoated printing stocks of different weights and surface characteristics, designed for various commercial and publishing applications.

In some cases, the final printed surface simply cannot be simulated by an inkjet-capable substrate. For example, metal and glass surfaces, often used in consumer packaging, can only be inkjet proofed by laminating a printed transfer film onto the metal or glass surface.

Inkjet proofing developers have also worked with device manufacturers to develop new inks for proofing on other high-end devices, such as Roland's VersaCAMM wide-format printer.

Extended-gamut or white inks from the inkjet device manufacturer may be used for proofing. However, in other applications—notably packaging—new ink formulations may be required.

Monitor-Based or "Soft" Contract Proofing

Physical inkjet proofs are cost-effective for many applications. However, they represent one of the few remaining physical steps in a mostly digital process. Thus, for an increasing number of print workflows, the review process takes place online and **on screen**, allowing multiple parties to annotate, comment, and digitally approve the job for printing.

Computer displays of PDFs or other digital files are adequate for content proofing—where color quality is not a critical factor. Monitors themselves have also improved. Modern LCD and OLED display technologies are capable of much greater color accuracy than their cathode ray tube predecessors. However, for contract quality soft proofing, just having a good color monitor is not enough.

The difficulty in soft proofing is partly due to the use of different color models, as discussed earlier in the chapter. However, it is also due to device consistency and changes in ambient lighting conditions. Placing a color display near a window, or in a colorfully painted room, will affect how the user perceives color on screen.

Soft proofing requires a high-end monitor that can be precisely calibrated, and that does not "drift" significantly over time. It also requires a controlled physical environment, e.g. neutral gray surroundings, and consistent ambient lighting conditions, to prevent undue interference from outdoor light changes.

Several vendors have created viable soft proofing environments—often integrated with online, secure review and approval workflow systems. Soft proofing systems usually require the use of a spectrophotometer to calibrate the device, keep it consistent, and create reliable profiles for designer use.

Packaging is a segment of the printing industry where soft proofing and design are converging. Unlike other forms of printing, packaging is always a three-dimensional challenge. Visualizing the finished result is a very different process from that of publication or commercial printing.

In the past, approval for a package design—including color choices—required a physical prototype. However, the high cost of production often limited the use of prototypes to high-end packaging jobs. More recently, advances in 3D modelling and display rendering have made "virtual reality" package proofing a cost-effective alternative.

Vendors of 3D packaging design software, with remote review/approval features and—in some cases—contract-quality color, include Esko Graphics, Creative Edge Software, and others.

As with inkjet proofing, soft proofing is a rapidly growing and changing field. As both become normal for graphic communicators, they will dispel the mystery, and the frustration, of dealing with color.

3D package design software is increasingly being used for soft proofing, allowing remote decision makers to approve both design elements and color choices. (Image courtesy of Creative Edge Software.)

7
Paper, Ink, and Toner

Paper, Ink, and Toner

Chapter Preview

Essential Qualities

Paper Manufacturing and Characteristics

Specifying Paper for Printing Applications

Paper and Sustainability

Ink Manufacturing and Characteristics

Ink and Toner for Digital Presses

Overview

Since its invention in A.D. 105, the material we know as paper has become the dominant substrate or surface on which images and text are reproduced. It was easier to manufacture than its predecessors: clay and papyrus. Its characteristics include durability, thinness, strength, weight, longevity, and the ability to receive an image from many different imaging and printing processes. Paper is ideal for graphic communication on any scale.

Other materials can serve as printing substrates, including plastics, foil, vinyl, glass, and metal—used in packaging and signage applications, and covered elsewhere in the book. There are also modern substitutes for fiber-based paper, such as synthetic and electronic substrates. However, for the vast majority of printing applications, traditional paper is still the primary means of mass-producing images and text.

This chapter will introduce you to how paper is made, its nature and properties, and how designers and print buyers can more effectively use this versatile medium. It also addresses a misconception about paper, namely its environmental sustainability.

We will also cover the colored substances we use to print on paper—ink and toner. Often customized for specific types of printing, or even specific devices, these substances are chemical combinations of pigments and a "carrier" designed to apply them to the printed surface.

Paper and ink have evolved significantly, paralleling changes in the print technologies that use them. Originally the product of manual craftsmanship, paper today is precisely formulated for highly-automated, high-volume manufacturing. At the same time, paper provides the same unique, tactile experience it has for almost two millennia.

Image courtesy of Sappi Global. (Photograph by Paul Close.)

The Essential Qualities of Paper

Graphic communication requires a **physical medium** with which visual storytellers can express ideas to the largest possible audience and at the lowest possible cost. These storytellers (writers, editors, illustrators, photographers, and designers) typically work on behalf of clients or employers and depend on an array of skilled partners to deliver their message. Their communication medium must have a balance of several important conditions:

Legibility includes a combination of visual qualities, such as *reflectivity* (light bouncing off the surface rather than passing through it) and *contrast* (the ease of distinguishing elements placed on the surface). Secondary factors include its color and texture, insofar as they affect how a placed image is perceived.

Durability and **Permanence** are important requirements. The medium must be able to survive normal use for a reasonable period of time. While that time interval varies widely—and no medium can be expected to last forever—it should last at least as long as people need it.

Environmental Responsibility is a complex, and relatively recent requirement. Manufacturing, use, and disposal of the medium must be considered to enable future generations' ability to safely and economically use the planet's resources.

Transportability involves the total weight and volume of the medium as well as its ability to withstand various shipping and storing conditions such as temperature and humidity. A good communication medium must be reasonably portable and stable in transit.

Aesthetic Appeal is a significant consideration in determining a medium's effectiveness beyond its purely utilitarian requirements. An exceptional communication medium connects on multiple levels, not only visual, but also tactile, and even audible and olfactory.

Affordability is the key to all the considerations previoiusly listed. As a rule, paper represents a significant percentage of printing costs—as high as 50% or more. While pricing details are outside the scope of this book, graphic communicators must develop an understanding of cost factors relative to the qualities of their preferred medium.

As noted in Chapter 2, there were some less than successful attempts to create such a medium before the 2nd Century invention of paper. Stone, clay, and bark were severely limited in many ways. Even papyrus, a relatively superior woven medium, did not meet every requirement.

For more than 19 centuries, fiber-based paper has proven to meet all these requirements. Alternatives—including synthetic substrates, plastics for packaging, and digital media—have not displaced paper significantly as the dominant medium for graphic communication.

Paper Manufacturing

Despite the dramatic increases in papermaking efficiency, the actual process is essentially the same as it has been for many centuries.

Paper typically starts as a slurry of fibrous material, separated from the non-cellulose parts of the original source, beaten or ground into a pulp, and mixed with water, additives, and colorants.

The fiber is usually plant-based, typically from wood and other sources high in cellulose. Other plant fibers may include cotton or hemp, as well as recycled fiber from converting processes, or from post-consumer sources.

Although fine printing papers often contain a certain amount of post-consumer fiber, high ratios of recycled content are used mainly for lower grades of paper or cardboard, because the cellulose fibers from recycled sources are shorter and less durable than those derived from virgin wood.

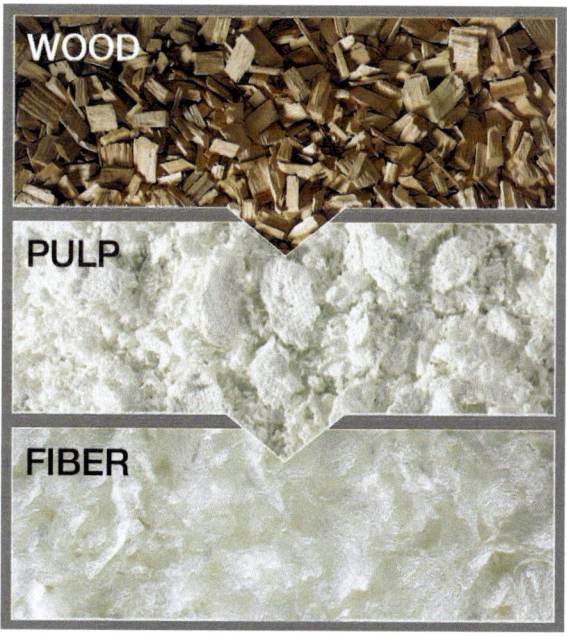

Using a Fourdrinier machine (named after its 19th Century inventor), papermaking consists of removing water from the pulp slurry, through gravity assisted drainage, along with pressure and heat applied to a continuous stream or "web" of material. The process causes the fibers to align in mostly the same direction, which gives the sheet its strength and, as a consequence, makes folding in one direction easier than the other.

Introduction to Graphic Communication | 117

During the water extraction process, the paper's texture is formed by synthetic forming fabrics or "dandy" rollers. Watermarks—as the name would suggest—are also created at this stage.

Once formed, more water is removed in a pressing section using high pressure hydraulic loaded nips where the sheet is supported with synthetic felts. Finally, the remaining water is evaporated as the web passes over a series of large, steam injected, drying cylinders.

Depending on how the paper is to be used, different coatings or starches may be applied to its surface. The paper is then smoothed by passing through high-presure rollers and heating elements—a process called calendering. Finally, the finished paper is either wound onto large rolls or cut into individual sheets. Paper rolls and "sheeted" paper are carefully packaged to preserve moisture integrity and press-readiness during shipment.

Paper manufacturing involves the use of large quantities of water. In fact, when paper is in its forming stages, it is made of up to 99 percent water. This is why paper mills are typically located near rivers and other fresh water sources. Until relatively recently, wastewater from Fourdrinier machines—often containing pulp contaminants and chemical additives—was dumped into the original water source. Environmental regulations—combined with increasing consumer pressure towards environmental responsibility—have greatly reduced this practice. Paper mills today typically use less hazardous additives. They also treat and cool the plant's waste water before returning it to the source.

Safety is another concern. Fourdrinier machines, often as long as a football field or longer, process a continuous stream of material at 40-60 miles per hour (64-97 kilometers per hour). It passes through a complex series of rollers, heating/drying units, coaters, and cutting blades. Anyone who flinches at the thought of a paper cut will appreciate the dangers of this fast-moving web.

As with environmental issues, mills and government agencies have begun to address safety concerns. This has not been without difficulty. Papermaking machines are huge, costing hundreds of millions of dollars and often take years to build. They are not easily replaced, although mills have taken significant steps to retrofit them for safety, product quality improvements, and manufacturing efficiencies.

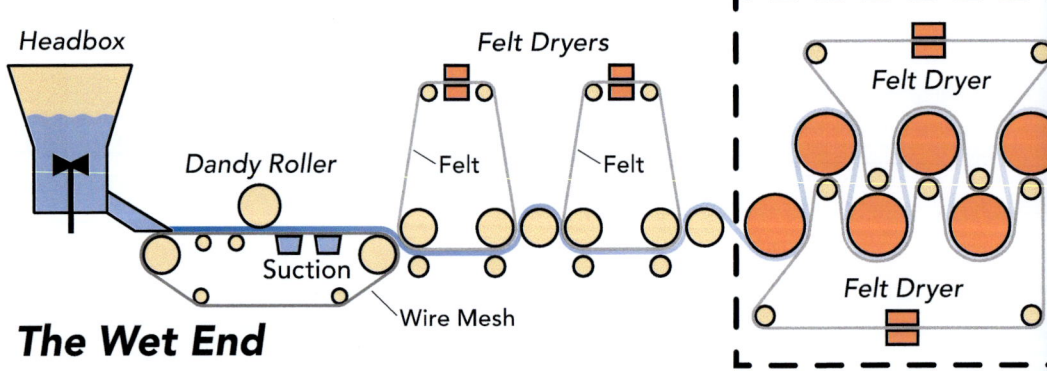

The Wet End

Although modern papermaking boasts a high degree of automation, it is not unusual to have a sizeable workforce standing by to deal with emergencies. A "web break" in a miles-long, fast-moving stream of paper is a serious matter. However, the coordinated response to such an event can reduce the potential down-time from many hours to a matter of minutes.

Paper Characteristics

Conventional, fiber-based paper can vary widely by many factors, the combinations of which result in a bewildering array of printing substrates—from premium white paper to cardboard and linerboard. (Other paper categories, such as tissue and non-woven materials fall outside the printing discussion.)

The characteristics of paper include size (sheet dimensions and thickness), weight, bulk, stiffness, tensile strength, and opacity. They also include surface characteristics such as color, brightness, whiteness, smoothness, gloss, and printability.

For centuries, sheet sizes were determined and named according to common usage or convention of the time. Names such as letter, ledger, tabloid, broadsheet, foolscap, and quarto, each had an historic meaning, and were associated (more or less) with specific measurements. However, as graphic communication became more industrialized, and more global, traditional names gave way to a more standardized approach.

Outside the U.S. and Canada, paper size has been codified by the International Standards Organization (ISO). Although there are several usage-based subsets within the ISO list of paper sizes, the most common in general for commercial printing are ISO A and ISO B.

ISO paper sizes typically have the same aspect ratio (1 to 1.41), from the largest (A0 or B0) to smaller sizes such as A4, used in office printing.

Paper sizes in the U.S. and Canada do not follow the ISO system, although U.S. printers often identify presses according

SELECTED ISO PAPER SIZES

	Millimeters		Inches	
	Width	Height	Width	Height
A0	841	1189	33.11	46.81
A1	594	841	23.39	33.11
A2	420	594	16.54	23.39
A3	297	420	11.69	16.54
A4	210	297	8.27	11.69
B0	1000	1414	39.37	55.67
B1	707	1000	27.83	39.37
B2	500	707	19.69	27.83
B3	353	500	13.90	19.69
B4	250	353	9.84	13.90

Smaller sizes (A5-A8 and B5-B10) not shown.

SELECTED ANSI PAPER SIZES

	Millimeters		Inches	
	Width	Height	Width	Height
ANSI A*	215.9	279.4	8.5	11
ANSI B**	279.4	431.8	11	17
ANSI C	559	432	22	17
ANSI D	864	559	34	22
ANSI E	1118	864	44	34

* US Letter   ** Ledger or Tabloid

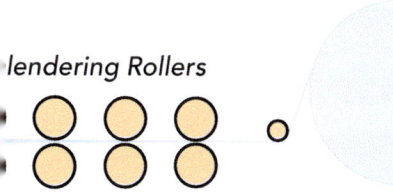

Calendering Rollers

The Dry End

to their optimal ISO sheet size. The U.S. system was published by the American National Standards Institute (ANSI) in 1995.

Paper thickness or **caliper** is also measured differently in the U.S. than it is in the rest of the world, which uses the metric system. Typical paper caliper can be between 0.0025 and 0.012 inches in thickness (or 0.06 and 0.30 millimeters), although cardboard and corrugate stock can be thicker.

A common way to describe different paper types is by weight for a given surface area. Outside the U.S., the measurement is expressed in grams per square meter—g/m^2 or simply gsm.

In the U.S., paper is defined by **basis weight**, namely the weight of 500 sheets of a particular size. This is complicated by the fact that the sheets for a particular kind of paper are of different sizes. For example, book stock, such as the page you are reading now, is calculated on the basis of a 25x38 inch sheet (3,300 square feet), while cover stock is calculated from a 20x26 inch sheet (3,000 square feet). Thus, the basis weight of 80-pound (80#) book paper is less than that of 80# cover, or $118 g/m^2$ versus $216 g/m^2$.

Fortunately, paper weight tables are very common in most printing environments, and in numerous online references and calculators.

Many properties of a sheet of paper are determined primarily by its blended fiber content—hardwood, softwood, alternative source fibers, and recycled matter. The latter can be unprinted converting scrap or post-consumer sources, which are de-inked and otherwise processed.

Fiber characteristics vary. Softwoods such as pine and spruce produce long, flexible fibers that give paper greater strength and durability. Hardwoods like aspen, maple, and birch produce shorter, stiffer fibers. Because of its processing requirements, recycled fiber is shorter still, which can limit the use of the resulting paper or cardboard.

The fiber component of paper is analogous to a skeleton. It gives paper its bulk, stiffness, and opacity. So, then, a paper's

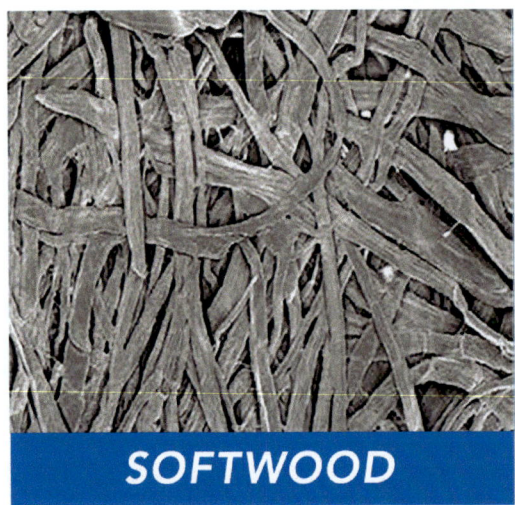

Wood fiber used in paper, at extreme magnification. (Image courtesy of Sappi North America.)

coating is analogous to skin. It smooths the paper surface, provides a base for printed detail, delivers color accuracy, and impacts printed ink gloss.

About 80 percent of coatings are pigments that enhance the paper's characteristics, such as clay for smoothness, calcium carbonate for opacity and brightness, and titanium dioxide for brightness. The remainder are binders such as starch or latex and various additives or dyes.

Optical brighteners are chemical additives designed to enhance image brightness and contrast for high-end printing applications. One side effect of optical brighteners in paper is their tendency to create a blue cast, which may in turn create undesirable color shifts in certain lighting conditions. As discussed in Chapter 6, this can be remedied with the proper use of color management.

The primary reason for coating paper is to enhance image clarity. Uncoated paper reflects light unevenly and allows ink to spread into the fiber. This results in a blurring or distortion of text and, more importantly, less sharpness, definition, and color separation in halftone images and tints. Coating of different types and surfaces (matte, satin, and gloss) correct these problems and provide sharper image results.

Coatings also allow for different types of use. Matte coating more easily allows handwriting and provides some of the softer look and feel of generally associated with uncoated stock, while providing the print quality benefits of coated. Silk or dull coating provides moderate reflectivity, and is good for fine art, fabric, and skin tone detail. Gloss coating

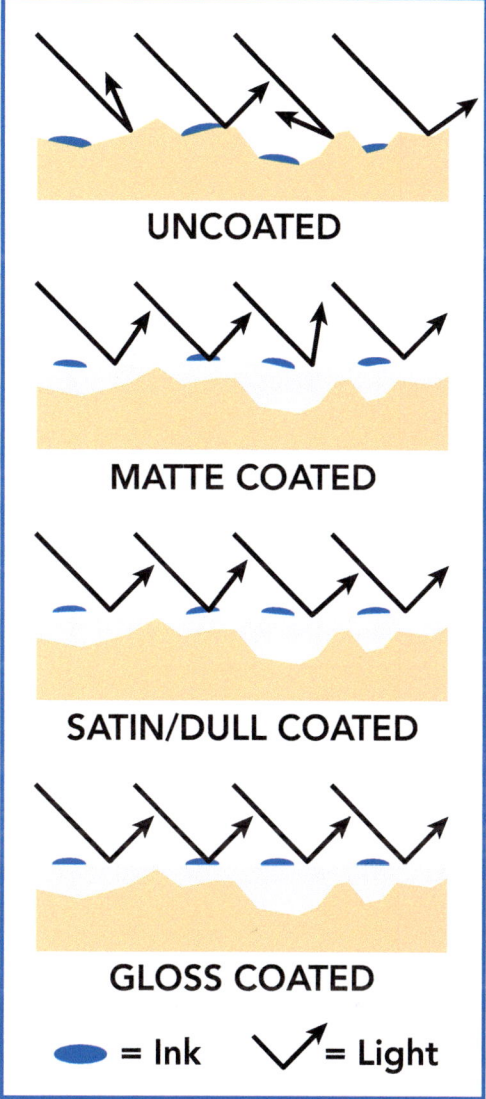

provides the highest potential for accurate dot reproduction and is best for reproducing images of hard or shiny objects.

Specifying Paper for Printing Applications

Specifying paper for a particular job can be challenging for designers and print buyers. The variables include several important factors:

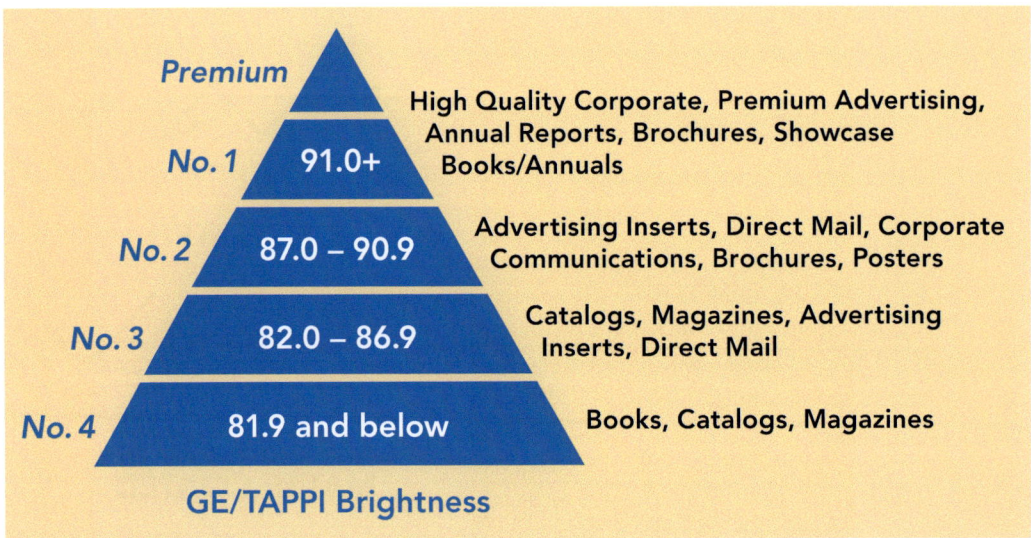

Using the North American model, paper is classified by different levels of brightness. Generally speaking, each level indicates a common use for specific applications.

The printing process itself is a major consideration. Many papers are optimized for a particular print processes such as offset or gravure, sheetfed or web, so both cost and printability considerations must go into the paper selection process. Papers designed specifically for certain digital print processes may run less effectively—or not at all—on conventional offset presses, or even on other digital devices.

End use is always determinative. Reading a book involves a different set of conditions and assumptions than filling out a form, following a posted sign, or browsing a store circular. Good designers must choose the right paper to meet these complex conditions.

Color and surface properties are highly relevant when choosing a paper, as are **basis weight and caliper**, as previoiusly noted. The physical and aesthetic attributes of a printed page—color, texture, reflectivity, "heft," and even smell—affect the overall perception of the printed message. Research in the field of haptics (the science of touch) has shown that the user experience with a printed page is superior in several ways to that of an electronic screen—even for those more accustomed to digital device use. Similar studies have shown that a paper of higher quality typically elicits a more favorable response than thinner or otherwise less substantial stock.

Brightness is often used to determine which paper is suitable for different publishing purposes. Technically, paper brightness is amount of reflectance of a specific wavelength (457 nanometers) of blue light, with 100 percent being the theoretical highest level (according to North American measurement standards). However, there is no universally-accepted standard for paper brightness. North America follows the GE/TAPPI (T 452) specification, while other parts of the world often use the *ISO* (ISO 2470-1) and *D65* (ISO 2470-2) brightness measurement methodologies. Due to their optical brightener content, the brightness of some papers will exceed 100 when measured by ISO and D65 standards.

Because the different methodologies do not directly correlate, it is sometimes difficult to compare the brightness of domestic and imported papers.

Not all printing applications adhere to these classifications. It is common to refer to "a number 3 sheet" as typical for magazines, although luxury consumer titles may opt for a brighter sheet.

Alternative Substrates

Partially in response to environmental concerns related to forestry, some have explored alternatives to paper made from wood fiber. An early example of synthetic paper manufacturing was Dupont's invention of Tyvek, a synthetic material resembling conventional paper but made with high-density polyethylene fibers—in other words, plastic.

There are two varieties of alternative substrates, sometimes referred to as "synthetic paper": one produced through an extrusion process such as Tyvek and the other produced with a coating blade.

The extrusion process allows the plastic material to flow through a screen-like device made up of very small holes that converts the plastic into fibers resembling those of wood pulp. The fiber structure can be left visible or coated.

The alternative process, developed by Yupo Corporation, uses a blade to spread the plastic into a continuous sheet of varying thickness. The coating blade process does not have the fiber structure of the extrusion process.

There is also a substrate made from stone (ViaStone). This is a unique material is made from natural stone, inorganic mineral powder, and non-toxic resins. It is particularly suited for inkjet printing.

Alternative substrates several special conditions and properties, which mirror the discussion of conventional paper:

Printability. While synthetic paper may have the look and feel of traditional paper, its printing requirements differ. One important difference that impacts printability is synthetic paper's lack of absorptivity, also known as "ink holdout."

Ink printed on traditional paper dries to a certain extent by a combination of absorption, oxidation, and heat. However, on a synthetic substrate, ink requires a more sophisticated drying apparatus. These include ultraviolet (UV), infrared (IR), or electron beam (EB) dryers, which dry/cure the ink instantaneously. Without these measures, there is insufficient time for an oxidation drying process to take place, and the sheets will stick together.

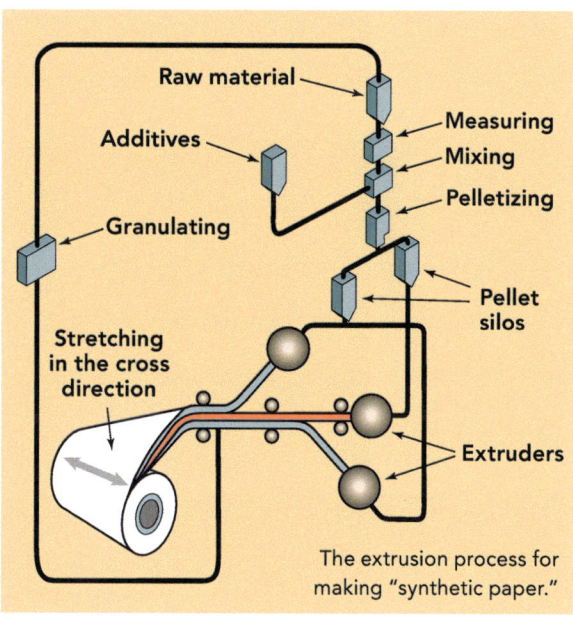

The extrusion process for making "synthetic paper."

Longevity. Librarians and archivists appreciate the promise of synthetic paper because of the potential for deterioration with pulp-based paper. While some books have survived for over 500 years, a typical publication made of traditional paper deteriorates much sooner due to the acid content of chemicals used in papermaking. Although significant efforts have been made to reduce or eliminate the acidity of conventional paper, the appeal of synthetic paper remains strong.

Resistance to moisture and tearing is particularly appealing in some use cases, such as for children's books and manuals, or documentation used in outdoor situations. Synthetic paper is more durable and long lasting, as it does not easily tear and can be washed.

Production Cost. As previously noted, conventional papermaking equipment require many years and hundreds of millions of dollars to build. Synthetic substrate-maufacturing equipment can cost less.

However, not all printing applications require or even benefit from alternative substrates. For many, the cost benefit and versatility of traditional paper still outweighs the appeal of its synthetic counterpart.

Paper and Sustainability

In the first edition of this book, contributing author Don Carli presented a compelling case for sustainability in print as part of a broader trend among corporations concerned about environmental and social responsibility. That treatise is still available online.

Sustainability remains a major concern for graphic communication. Life cycle studies have shown that paper manufacturing is the biggest contributor to the life cycle of printed materials. Because paper manufacturing uses very visible natural resources (trees and water), there is an understandable demand from consumers and the companies that serve them for paper that is produced and used in a sustainable manner.

FSC® certification assures consumers that paper has been derived from sustainably managed forests.

Countering Common Misconceptions

As digital media make inroads into traditional publishing and communication, a commonly-heard argument holds that "going digital saves trees" or other "green" messaging. The implication is that a shift from print to online media will reduce the use of paper and energy, and also prevent trees from being cut down for paper manufacturing. This is simplistic (at best) and ignores the fact that digital media require the continual use of electricity. The environmental cost of that energy is seldom calculated.

Contrary to conventional wisdom, paper is in fact one of the most potentially sustainable media for graphic communication, based on several facts:

Mills are taking extraordinary steps to limit the adverse effects of paper making, including on-site water treatment. (Image courtesy of Stora Enso.)

Fiber sourcing is now a well-established metric in the paper supply chain. By using fiber from sustainably managed forests—a truly renewable resource—paper mills can provide consistently high levels of paper quality without depriving future generations. Several organizations, including the Forest Stewardship Council® (FSC®) and the Sustainable Forestry Initiative (SFI), have developed forest management standards to help guide practices related to forest health including protection of soil and water quality as well as biodiversity. These programs also have chain of custody standards which provide a certification process to trace the origin of wood fiber, to ensure sustainable forest management and legal logging practices.

Recycled fiber usage is a consideration for some types of paper—mainly lower grades and corrugates—but is not a panacea, as noted earlier.

Water usage is a major consideration for paper mills, where a product begins as a slurry of up to 99 percent water. A majority of paper mills now responsibly manage water issues, including temperature and additive use, with environmental considerations clearly in mind.

Recyclability is an important aspect of the lifecycle of paper. While it is generally common knowledge that paper is recyclable, consumers are sometimes confused or forgetful, so it is recommended to include a logo or message reminding people to recycle printed pieces. Most paper mills are proactive in working with government and non-governmental bodies in determining how their diverse products can best be handled. Fortunately, the major component of paper—cellulose fiber—is a known quantity in the recycling equation.

By paying attention to the sourcing and responsible manufacturing practices of their suppliers, graphic communication professionals can be confident that theirs is a sustainable industry.

Ink and Toner

After paper, ink is the most expensive disposable commodity used in printing. In its broadest definition, ink consists of pigments (suspended particles) or dyes (liquid solutions) combined with an oil-based resin or aqueous "carrier" medium and other chemicals to maintain quality and stability—or aid in the drying process. Ink manufacturers often employ highly-specialized chemical processes on a vast, industrial scale. Originating in medieval times, modern ink creation has evolved far beyond its craft origins. It is a combination of manufacturing automation, digital color science, and creative research.

Ink Manufacturing

The types of ink are as diverse as the many printing processes that use them (see Chapter 8). Offset lithography, gravure, flexography, and screen-printing each have different requirements, including the formulation of ink. Digital presses also have their own unique requirements, as will be covered later.

In brief, the printing ink-making process begins with the selection of pigment particles that provide a particular color. Black ink uses carbon black—a substance made from the partial combustion of petroleum products. White ink—or lighter shades of colored inks—use titanium dioxide. Other inks, including process cyan, magenta, and yellow, and custom colors and metallic inks, each use chemical solids developed by the manufacturer.

The pigment particles are milled and separated to a uniform consistency and mixed with a varnish resin. Additional chemicals are added, and the results routinely tested for their spectral color characteristics, as well as drying and other press-critical requirements. After all processing is complete, the resulting substance is packaged and stored in conditions designed to preserve its printing qualities.

This simplified process contains many variables, including the substitution of non-petroleum-based carriers such as soy-based oils and chemicals with less toxicity and a more sustainable environmental impact. Ink manufacturers are well aware of environmental issues and have made great strides in developing more eco-friendly formulations, in most cases without sacrificing quality or printability.

Ink Properties

Printing ink obviously has **optical properties**, involving its capacity for absorbing or reflecting light of specific wavelengths. Whether those properties involve process color (CMYK) or specific spot or custom color, the science of color management, covered in Chapter 6, is now able to resolve differences in ink formulation that once made color reproduction difficult. In particular, the Color Exchange Format (CxF) has enabled ink manufacturers to create more precise and consistent formulations.

Related to color behavior is an ink's **opacity**—the degree to which it allows or prevents light transmission—and its **permanence**. Over time, an ink may not retain its original color strength and brightness when exposed to light. Depending on the application, a different ink formulation may be needed to prevent or minimize this fading.

Another physical property of ink is **wettability**—its ability to resist bleeding when exposed to water.

Ink **drying** occurs by a combination of methods, including (among others) oxidation, absorption, and heat-induced evaporation. Whatever drying method is employed, the optimum combination of ink and paper should be used to prevent ink chalking or "rubbing"—that unpleasant transfer of incompletely dried ink to one's fingers. Other problems associated with incomplete ink drying include show-through (common with thinner substrates) and set-off (the appearance of a printed image—reversed—on the sheet above).

On press, a number of ink properties can affect how well the press performs, including an ink's **body** or consistency and its **viscosity**. Another ink property is "**tack**"—its stickiness, which affects how easily a printed sheet can be separated from the actual printing surface, such as the rubber blanket on an offset lithographic press.

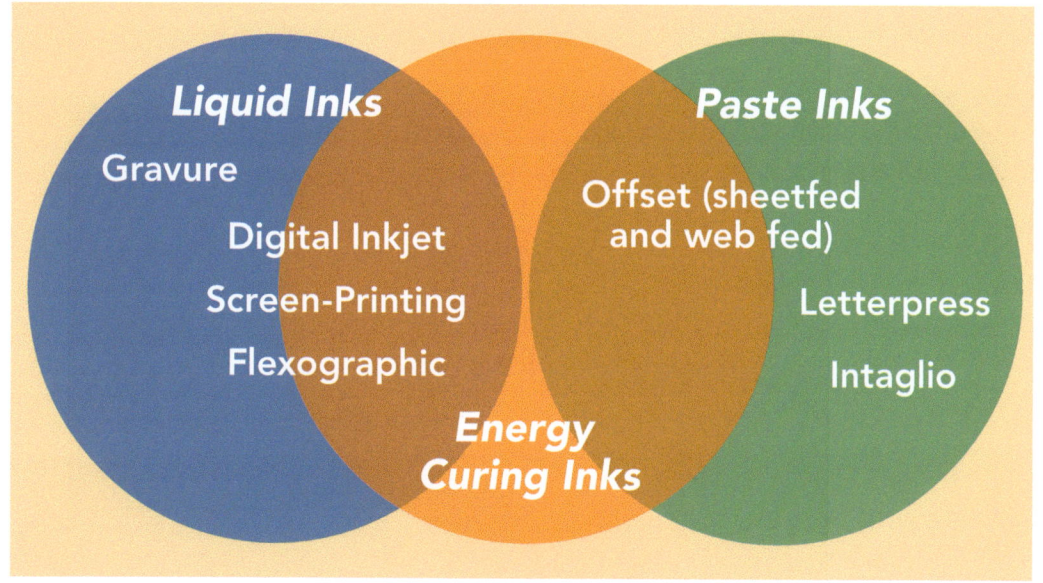

Digital Inkjet Inks

The rise of digital printing has transformed many aspects of graphic communication, as covered in the next chapter. In particular, inkjet presses are well positioned to replace conventional presses in a growing number of print applications.

However, the inks used in these devices are substantially different from those previoiusly described. They are primarily aqueous in nature, that reduces their potential for adverse environmental impact. They also use a fundamentally different path from the device to the paper surface. Instead of being squeezed onto the paper from a plate or cylinder, the inks are forced through a microscopic nozzle, at precisely controlled intervals, in a non-impact environment.

Inkjet inks are almost always formulated by the device manufacturers, who intimately understand the mechanics required to create a printed image. Although this precludes users from "shopping" for third-party inks, it also provides simpler solutions for common print reproduction issues.

One such issue involves drying. At high speeds typical of high-end inkjet presses, droplets of ink must adhere to the paper surface and dry almost instantaneously, to avoid smearing and other undesirable outcomes. Another involves machine efficiency. Digital inkjet presses typically use large cartridges that can be easily replaced without disrupting production, although it is theoretically possible for a third party to develop cartridges for different press manufacturers, it is generally more efficient for the original equipment manufacturer (OEM) to do so.

Inkjet press manufacturers very often work closely with paper manufacturers to ensure efficient, high-speed color reproduction. The number and variety of affordable, inkjet-compatible printing papers is expected to increase dramatically to meet the needs of designers and print buyers.

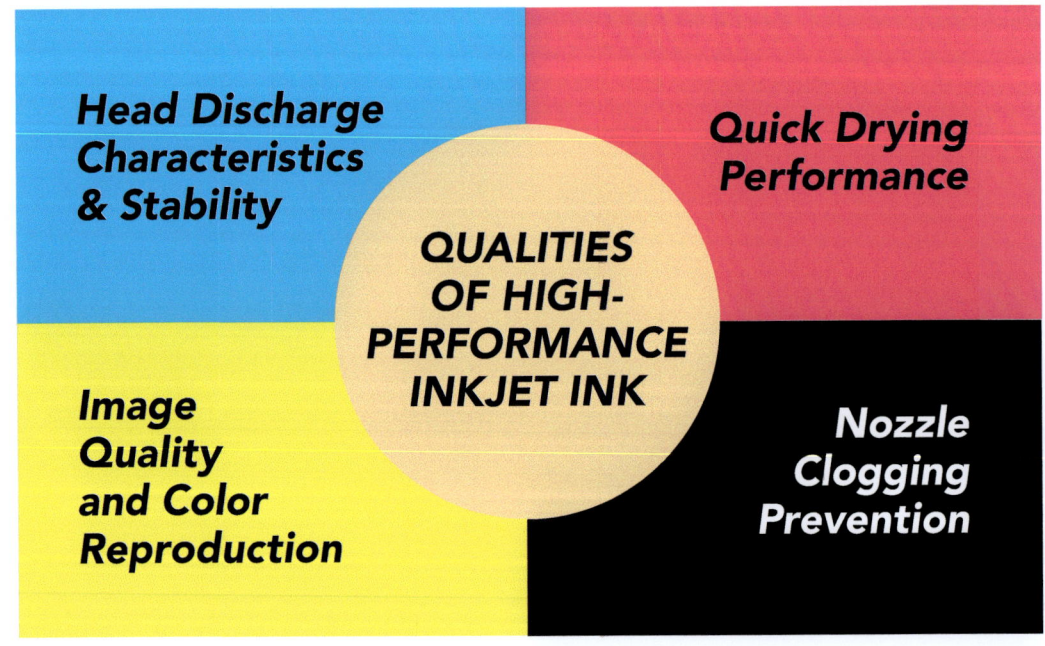

Digital Electrophotographic (EP) Toners

The other type of digital press (also covered in the next chapter) is based on electrophotography, often referred to as "xerography." Instead of inks, these devices use electrically charged particles—toner—to produce an image, which is eventually fused onto the paper.

Toner manufacturing is also a complex chemical and mechanical process. The source material is typically a polymer, formulated for specific color qualities and for a specific melting point—namely the temperature generated by the printing mechanism. The material begins as a solid slab, which is broken apart or pelletized into fine particles—approximately 8-10 microns in diameter. Finer sizes require a chemical process to "grow" the toner particles from molecular reagents.

Pelletized toner particles can also be electromagnetically sorted, to achieve a uniform size. Larger, separated particles are typically melted and re-used to create new source material. Uniformity is the key. The rule of thumb for toner is that smaller, more uniform particles result in more accurate color reproduction.

Similar to ink for inkjet devices, toner is manufactured primarily by the OEM, where it is made according to the exact requirements of the EP device. Using third-party toner in a device is almost always ill-advised. The money saved will invariably be offset by repair or replacement costs, since variations in toner particles can significantly affect performance or the lifespan of the device.

Toner-based device OEMs also work closely with paper manufacturers, to ensure that their toner particles adhere properly—and permanently—to the sheet. These devices do not content with drying issues, and so are compatible with a wider variety of paper stocks. However, most EP presses are more limited in the size of paper they can handle.

In graphic communication, whatever paper or printing method is used, properly made inks and toners are what create the image a user sees. Think of ink and toner as the text of the message, and of paper (or other substrate) as the context.

Introduction to Graphic Communication | 129

8

Printing Processes

Printing Processes

Chapter Preview

The Printing Press
Attributes of Print
Traditional Print Processes
Combination Printing
Digital Printing Benefits
Digital Print Engines
Wide Format Devices and Digital Printing Presses
Print Economics and the Future of Print

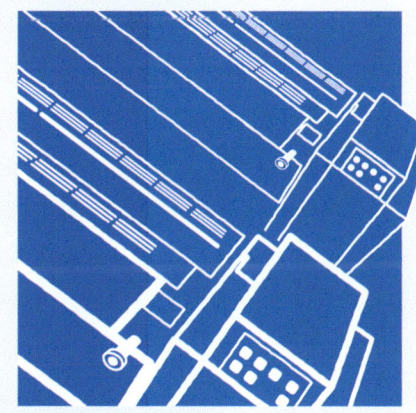

Overview

Physical printing involves colored inks, toners, or other colorants repeatedly applied to sheets of paper, plastic, metal, or other materials. Conventional print processes range from traditional letterpress (raised surface) and gravure (engraved surface) to offset lithography, flexography, and screen-printing. Digital printing includes both inkjet and electrophotographic (EP) processes.

This chapter covers the mechanical principles of putting ink or toner onto a substrate. Armed with this basic technical knowledge, you will be able to make better decisions when planning, designing, and managing your print projects.

A Heidelberg sheetfed offset press and a manroland web offset press, both with multiple, in-line printing stations or units.

The Printing Press

The press is the basic unit of the entire printing process. It is a precision instrument—far evolved from Gutenberg's modified wine press. Although a press is typically the largest and heaviest piece of equipment used in printing, it is highly controllable—to thousandths of a degree—on matters such as cylinder pressure, color balance, and image positioning. Often operating at high speeds, with computerized automation, and working in concert with many other systems, a press is a formidable piece of equipment.

Presses are configured as either **sheetfed** (individual sheets of cut paper) or **web** (continuous rolls of paper to be cut after printing). They are configured with printing stations or **units**—each one printing a different color or providing other applications such as special coatings or varnishes, to enhance the look of the finished piece.

A typical configuration is a four-color press, with each of the units printing one of the four process colors (CMYK), as discussed in Chapter 6. One-color and two-color presses can suffice for simpler applications, while five- and six-color presses are common in commercial printing environments. Hi-fidelity or HiFi printing requires at least six units—typically adding orange and green ink to the normal CMYK process inks. Presses are also manufactured with more than six units for specialty printing purposes, especially where custom or "spot" color inks or varnishes are required.

Multiple units allow the press to print more than one color on a substrate in a single pass through the press. (It is possible to print a four-color job on a one-unit press, but that requires a separate set-up—and wash-up—for each pass.)

Another important configuration is **the combination press**—those that incorporate more than one printing process. For example, some presses use both flexographic and gravure printing processes to create product packaging. Another common combination press uses a combination of offset lithography and inkjet printing to create color publications or direct mail pieces with unique, customized information (such as a mailing address) printed on each piece.

Combination presses are also used in security printing, where documents must be produced that are difficult to copy or counterfeit. Scratch-off gaming or lottery tickets are produced on combination presses using many processes, including lithography, gravure, flexography, and inkjet printing. These presses sometimes have as many as sixteen units, because of the number of colors printed as well as the various layers of coatings needed to enhance security.

Every printing process is characterized by its image carrier or plate characteristics. They are also characterized by the formulation of their inks (or toners in the case of electrostatic or electrophotographic or EP printing) and of the substrate types they are capable of handling. For example, not every kind of paper can be used with certain digital presses; special coatings are required for the inks or toners to adhere properly.

Each process can be identified under magnification, usually 12x or greater, by someone skilled in the art of printing. Offset lithography has a different look than letterpress; gravure has a different look than flexography; inkjet printing has a different look than EP. Although the quality of print produced by these methods is constantly improving—making it harder to tell with the naked eye—important differences remain.

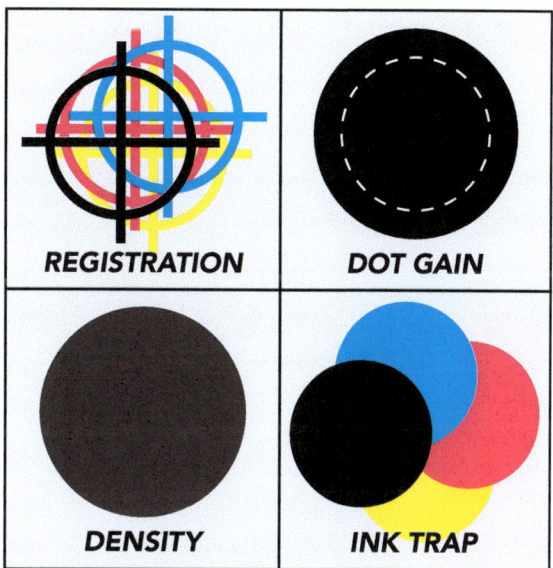

Common printing attributes include registration, dot gain, ink density, and trapping.

Important Print Attributes

Regardless of what technique is used, there are four important attributes of a printed color image. Each of these can have a significant impact on the quality of the work.

Registration is the relative positioning of the image on the substrate and the relative positioning of each ink layer over another. If the positioning of ink colors over each other is not accurate, the printed image will appear blurred.

Dot Gain is the extent of growth that takes place in the size of a halftone dot or screen tint dot from the printing plate to the printed sheet. All printing processes—with the exception of electrostatic or electrophotographic printing—are subject to dot gain. It occurs because the ink is squeezed onto the substrate under pressure. Because the ink is a liquid and has a thickness to it, there is a tendency

Introduction to Graphic Communication | 135

for the ink to spread under the pressure. In inkjet printing, the ink is squirted onto the substrate and the force of the ink spot on the substrate causes the dot or spot to grow. Dot gain does not occur in electrostatic printing because dry toner particles are used that do not grow. Dot gain influences the look of a printed image. It is expected on the printing press and can be controlled in platemaking or in building digital files for printing. For example, if a dot gain of 30 percent is expected in the magenta ink being printed using the lithographic process, there are ways of reducing the size of the halftone or screen tint dot on the printing plate that will print the magenta ink.

Density is the intensity or visual strength of the ink that influences the color quality of the final printed image. In full-color printing, standards have been established for ink density when all other printing press variables are properly controlled. In other words, the density standard or target for yellow, magenta, cyan, and black are different in full-color printing. A densitometer is used to measure the density of ink on a printed sheet. (A spectrophotometer, is used to measure the actual color wavelengths as well as other attributes of color.)

Trapping is the extent to which one film of ink sticks to another when printing one ink film over another. In four-color printing, the yellow, magenta, cyan, and black halftone dots must partially overprint each other to satisfactorily produce the resulting red, green, and blue colors needed in full-color printing. Ideally, 100 percent of an ink film will stick to the other. This often occurs when a wet ink film is applied to a dry one. However, in reality, because the inks are wet on a multicolor press, less than 100 percent of one ink film adheres to the other. If only 85 percent sticks or transfers, this is called 85 percent trapping. The degree to which trapping occurs influences the look of the final print. The percentage of trapping too is measured with a densitometer or a spectrophotometer.

The term trapping is also used to describe the creation of overlapping areas between two adjoining colors. The goal of these overlaps—created by specialized software like Luminous TrapWise and Scitex Full Auto Frame—was to avoid noticeable gaps caused by press or plate registration errors. However, that process is now handled in the RIP, rather than by a dedicated software application. The problem itself has diminished with recent improvements in press registration technology.

Other components of a press that impact the appearance of the final printed piece include plate-to-blanket squeeze pressures (on offset presses), ink film thickness, the balance of fountain solution (water) and ink, the pH

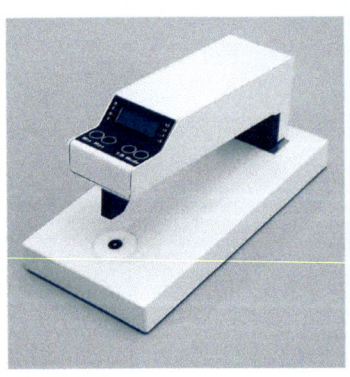

A densitometer measures the degree of darkness or light absorbance of a material—in this case a layer of ink or toner covering a substrate.

and conductivity of the fountain solution, roller settings and roller hardness, the tension of the substrate going through a web press, and more. Controlling these variables scientifically—and with less manual labor—has been the goal of modern press manufacturers for many decades.

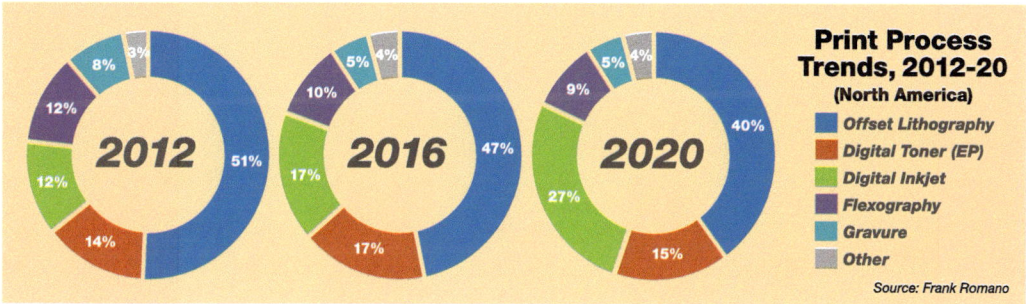

Traditional printing processes are being significantly displaced by digital. Shown here is the percentage of North American print sales (customer price per page impression), by printing process, 2012–2020.

Traditional Printing Processes

Although nearly all printing today has a digital component, there are five technologies known as conventional or traditional printing processes: **letterpress, offset lithography, gravure, flexography,** and **screen-printing.** With the exception of letterpress, all are represented by major press manufacturers, and occupy a substantial portion of the multibillion dollar printing industry.

Although traditional printing has long dominated the industry, it is displaced by digital processes described later in this chapter. This is especially true of the centuries-old letterpress process. Here is an overview of the traditional printing processes:

Letterpress

Letterpress was once the mainstay of printing, but it is largely obsolete today. Briefly, letterpress is **printing from the surface of a raised image.** It is also called relief printing, where the plate image is uniformly higher than the nonprinting areas. By far the oldest of the conventional processes, it has declined significantly in the past 50-60 years, as more efficient, economical, and higher quality printing processes emerged.

Less than three percent of all printing in the United States involves the letterpress process. Its demise is directly related to the amount of set-up time required; the heaviness, cost, and cumbersome nature of the plates (made of zinc, copper, lead, and photopolymers); and image resolution limitations.

In letterpress printing, ink is placed in an ink fountain and is then distributed onto mechanized ink rollers. The ink rollers apply the ink onto the raised image of the plate, and the plate transfers the image onto a substrate—usually paper. The image on the plate must be "wrong-reading" (a mirror image) so that the printed image on the substrate will be "right-reading."

Introduction to Graphic Communication | 137

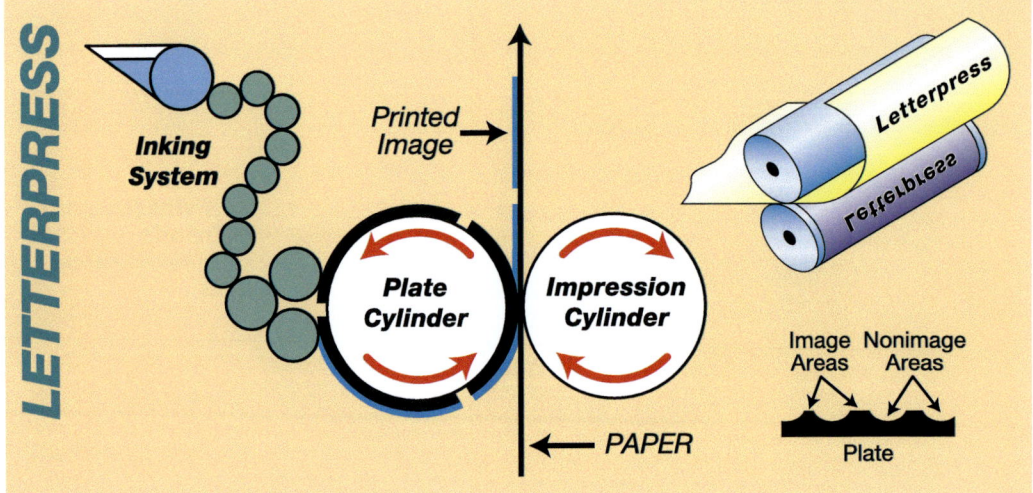

The principle of letterpress printing, a relief printing method where the image areas are raised above the nonimage areas. Rotary letterpress, showing the raised, wrong-reading type on the plate cylinder and the right-reading image on the paper.

Letterpress printing exerts variable amounts of pressure on the substrate depending on the size of the image elements being printed. The amount of pressure per square inch—or "squeeze"—is greater on some highlight dots than it is on larger shadow dots. Expensive, time-consuming adjustments must be made throughout the press run, to ensure that the impression pressure is accurate.

In traditional letterpress printing, letters were assembled into copy, explanatory cuts were placed nearby, line drawings were etched or engraved into plates—and all these were placed (composed) on a flat "stone" within a rigid frame called a chase, spaced appropriately with wooden blocks called furniture, and tightened or locked-up with toothed metal wedges called quoins.

In its heyday, letterpress was used to print a vast array of products and publications. However, with the exception of specialized art prints and other niches, letterpress has been relegated largely to the finishing processes of embossing, die-cutting, and foil stamping.

There are three types of letterpress printing presses: platen, flatbed, and rotary. The platen variety—often using a slow, handfed process—is the direct descendant of Gutenberg's modified wine press.

In a flatbed cylinder press, the plate is locked to a horizontal or vertical bed. That passes over an inking roller and then against the substrate. The substrate is passed around an impression cylinder on its way from the feed stack to the delivery stack. A single revolution of the cylinder moves over the bed, so that both the bed holding the substrate and cylinder moved up and down (or back and forth) in a reciprocating motion. Ink is supplied to the plate cylinder by an inking roller and an ink fountain.

Flatbed cylinder presses operated very slowly, with a production rate of not more than 5,000 impressions per hour. As a result, much of the printing formerly done on this type of press was moved to rotary letterpress or lithography.

Rotary letterpress required curved, image-carrying plates. Typically, these were created from the original, flat-surface plates, using molded plastic or rubber, and known as stereotype or electrotype plates. When printing on coated papers, rotary presses used heatset inks and were equipped with high-velocity, hot air dryers.

When letterpress was dominant, web-fed rotary presses were used primarily for printing newspapers. These were often "perfecting" presses—designed to print both sides of the web simultaneously. The web width was up to four pages across, although later presses printed up to six pages across on a 90-inch web. Although largely replaced by other processes today, rotary letterpresses were also used for long-run commercial jobs, packaging, book, and magazine printing.

Offset Lithography

Lithography is printing from a flat surface on which the image areas and non-image areas are on the same plane. The process is based on the principle that grease and water do not mix. The image and non-image areas are separated chemically in such a way that the image on the plate will accept greasy ink and the non-image areas will accept water and afterward reject ink.

On a typical lithographic press, there is an ink fountain and a water or dampening fountain. Ink is distributed from the ink fountain onto a set of ink rollers. Simultaneously, the water fountain distributes a dampening solution, primarily composed of water and as small percentage of chemicals that help the water desensitize the non-image areas. The combination is called fountain solution that is then applied to dampening rollers. The rollers dampen the plate before ink is applied to it. The water sticks to the non-image areas that were chemically treated to accept the water. The ink rollers then apply ink to the plate. Because the water on the non-image areas rejects the greasy ink, the ink will only stick to the image areas. The lithographic plate is typically made of aluminum, although other metals as well as paper and plastic can be used.

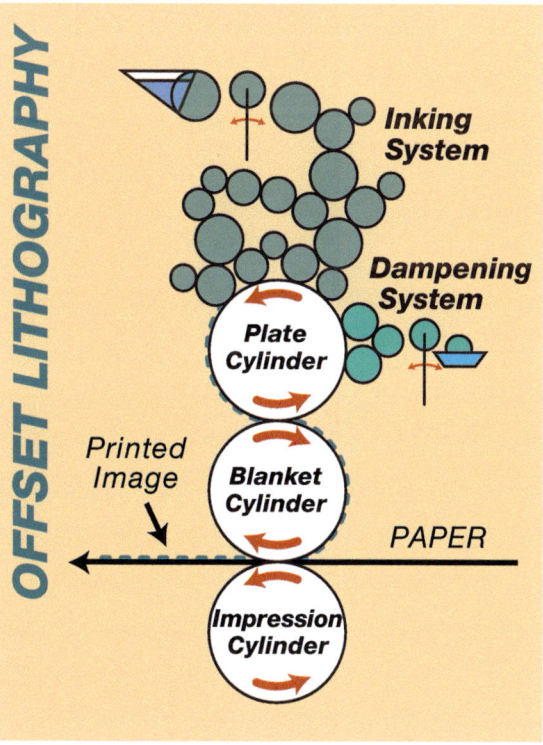

The principle of offset lithography, a planographic printing process where the image and nonimage areas are separated chemically in such a way that the image on the plate will accept greasy ink and the nonimage areas will accept water and afterward reject ink.

Introduction to Graphic Communication

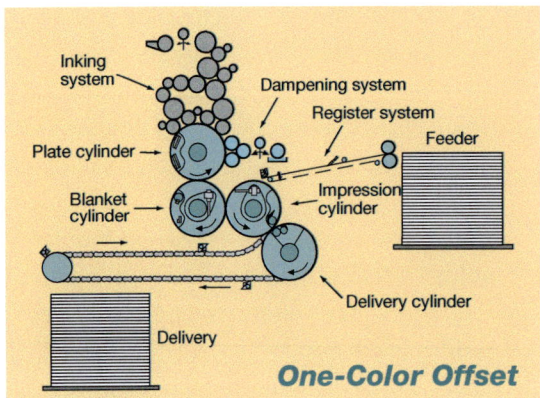

One-Color Offset

A single-color sheetfed lithographic press diagram showing the basic systems common to all such presses.

The inked images are then transferred to a synthetic rubber blanket that is wrapped around a cylinder that comes in contact with the plate cylinder. From the imaged blanket, the image is transferred to the substrate being printed.

The blanket performs three tasks. The first is to allow a right-reading image on the plate to become right-reading on the substrate. Without the blanket cylinder, the image would go from right-reading on the plate to wrong-reading on the substrate. The blanket's second function is to reduce the amount of fountain solution that reaches the substrate. When printing on paper, moisture absorbed by the paper causes paper distortion or dimensional instability. The third role of the blanket is to allow printing on a large variety of substrates—regardless of texture in most cases. The blanket allows for a certain degree of compressibility so, when printing on rough-textured substrates, the ink can be forced into the "valleys" of the paper.

Offset lithographic presses can be sheetfed or web. On the former, the substrate is fed into the press one sheet at a time at a very high speed, using highly specialized hardware for picking up and moving each sheet. Web-fed presses print on a continuous roll of substrate, or web, which is usually cut to size on the delivery end of the press.

Lithographic presses are classified as heatset or non-heatset, depending on the type of ink used and how it is dried. Heatset inks—typically used on high-quality, glossy stock—require special drying mechanisms on the press such as heating ovens, ultraviolet, infrared, or electron beam dryers. Non-heatset inks are typically dried via oxidation and absorption, and are used when the substrate is more porous, such as in newspaper printing.

Today, lithography is the most widely-used traditional printing process, used on a wide array of work, from simple, single-color to high-quality full-color work. It is well suited for printing text and illustrations in short to medium length runs of up to one million impressions. Approximately 50 percent of all printing in the United States is produced with the lithographic process, but its use is declining as digital processes improve in speed, capabilities, and quality.

Gravure & Engraving

The gravure process has its origins in the early seventeenth century when the intaglio printing process was developed to replace woodcuts in illustrating the best books of the time. In early intaglio printing, illustrations were etched on metal, inked, and pressed on paper.

Gravure, a common type of intaglio printing, makes use of the ability of ink to fill slight depressions on a polished metal plate. The process of gravure printing today consists of a printing cylinder, a rubber-covered impression roll, an ink fountain, a doctor blade, and a means of drying the ink.

In principle, gravure printing can be thought of as the opposite of letterpress printing. Where letterpress prints from a raised image, gravure prints from a recessed image. In gravure printing, the image area is beneath the plate surface and the non-image area is on the plate surface. A typical gravure plate is a large copper- or chrome-surfaced cylinder. Through chemical, electro-mechanical, or laser engraving processes, an image is etched or engraved onto the cylinder in the form of microscopic wells or cells.

Initially, ink in the gravure press is applied directly to the copper cylinder, not only filling the wells but also adhering to the surface of the cylinder. It is applied to both the image and non-image areas of the cylinder. However, a doctor blade made of hard rubber or plastic then passes over the cylinder and scrapes off ink from the non-image area on the surface. After this occurs, the substrate being printed comes in contact with the cylinder at high speed and under high pressure. As the paper is rapidly pulled off of the cylinder, capillary action pulls the ink out of the cylinder ink wells, which represent the image area, and the ink is transferred onto the substrate. This all occurs at a high rate of speed.

Gravure printing involves high costs, including the time required to prepare the plate cylinder. It is, therefore, economical for very long press runs where the cylinder does not have to be changed often. Typically, printing requiring tens of millions of impressions lends itself to the gravure process. Another unique aspect of gravure is its potential for producing high quality images. It allows for smooth tone transitions from highlights to midtones to shadows.

The dominant gravure printing process, referred to as rotogravure, employs web presses equipped with cylindrical plates as image carriers. A number of other types of gravure presses are currently in use. Rotary sheetfed gravure presses, though rare, are used when high-quality pictorial impressions are required. They find limited use, primarily in Europe. Intaglio plate printing presses are used in certain specialty applications such as printing currency and fine art. Offset gravure presses are used for printing substrates with irregular surfaces or on films and plastics.

Today, almost all gravure printing is done using engraved copper cylinders,

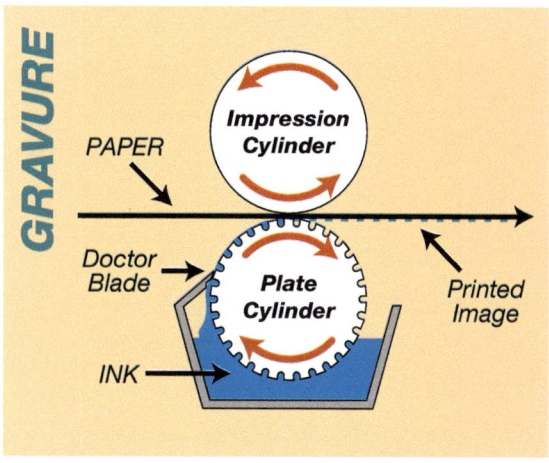

The principle of gravure, an intaglio printing process where the image areas are engraved, or recessed, below the nonimage areas.

Introduction to Graphic Communication | 141

A TR 10B gravure press.
(Courtesy Koenig & Bauer AG)

It is important to note that gravure is not the only form of intaglio printing. Traditional engraving is still part of graphic communication today, with more than 30 engravers in the United States alone.

Today's engraving methods involve etching an image onto a copper plate, which is then mounted on the press with a matching counter. The image is transferred from plate to paper resulting in a finely detailed, raised image on the paper's surface.

protected from wear by the application of a thin electroplate of chromium. Rotogravure cylinders vary widely in size, depending on the application. Publication press cylinders can be up to eight feet wide, while packaging press widths rarely exceed five feet. Specialized presses, for printing paper towels, can be 20 feet wide. The diameter of a gravure cylinder can be as large as three feet, or as small as three inches—for printing wood grains.

Gravure is a popular process for long-run, large volume publication and catalog printing, where image quality is vital. It is also used for package printing on non-paper or board substrates such as foils, plastics, cellophane, and other substrates having little or no absorption. It is also a popular process for printing on specialty items such as wall coverings and linoleum and for producing synthetic wood grains on pressure-sensitive substrates. Gravure represents approximately 12 to 13 percent of all printing today, but is gradually declining in use.

Flexography

Similar to letterpress printing, flexography involves printing from a raised image on the plate. The difference, however, is that the flexographic plate is typically made of synthetic rubber or a photopolymer material. Some of the harder flexographic photopolymer plates print relatively sharp and produce high-resolution images. However, the softer, synthetic

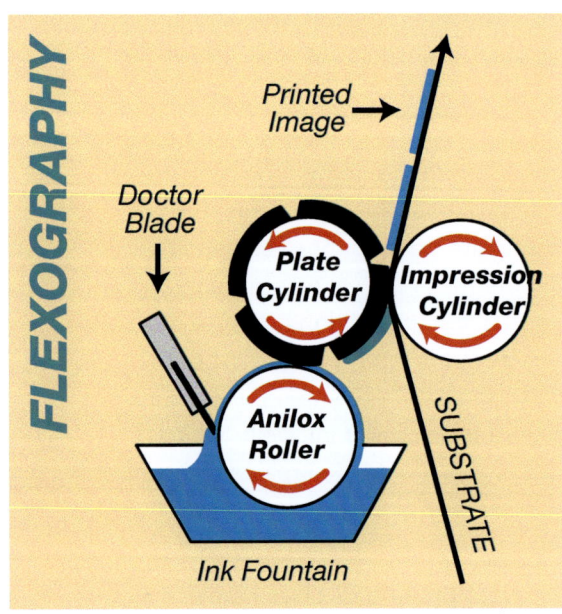

One variation of the flexographic process that uses an anilox roll and a doctor blade to apply ink to the raised surface of the printing plate.

rubber plates are not as suitable for high-quality printing and are typically used for long-run imaging requiring one or two flat colors where image sharpness is not of critical concern.

In the typical flexographic printing sequence, the substrate is fed into the press from a roll. The image is printed as the substrate is pulled through a series of stations or print units. Each print unit prints a single color. As with gravure and lithographic printing, various tones and shading are achieved by overlaying the four process ink colors—magenta, cyan, yellow, and black.

Flexographic printing begins with image and artwork preparation, in most cases comparable to the prepress process described in Chapter 5. Manual image assembly involving "mechanical" artwork and analog image photography is no longer frequently used, having been replaced with digital processes. Flexographic work, especially packaging, does require some skills that are distinct from typical print preparation, however. On-press color behavior is often different in offset printing and flexography, for example. Proofing is also different, due to the need to create three-dimensional proofs and prototypes, particularly for packaging.

Flexographic plates are relief plates that come in contact with the substrate being printed. The plates are attached to a roller or cylinder for ink application. They

Mark Andy LP 3000 flexographic, web label press (Courtesy Mark Andy, Inc.)

are made using three different processes: photomechanical, photochemical, and laser engraving.

There are five types of printing presses used for flexographic printing. These are the stack type, central impression cylinder (CIC), in-line, newspaper unit, and dedicated four-, five-, or six-color unit commercial publication presses. All five types employ a plate cylinder, a metering cylinder known as an anilox roll that applies ink to the plate, and an ink pan. Some presses use a third roller as a fountain roller and, in some cases, a doctor blade for improved ink distribution.

Flexographic inks are similar to packaging gravure printing inks in that they are fast drying and have a low viscosity. The inks are formulated to lie on the surface of non-absorbent substrates and solidify when solvents are removed by drying devices. Solvents are removed with heat unless UV-curable inks are used.

The technology of flexography has improved rapidly over the past decade, as has its quality. The process is popular for label printing, packaging, corrugated board printing, and for printing on non-paper substrates. These include cellophane, plastic, polyester, foils, folding cartons, paper bags, plastic bags, milk and beverage cartons, disposable cups and containers, adhesive tapes, envelopes, food wrappers, and other substrates with little or no ink absorption. In recent years flexography has also become popular for newspaper printing because the process lends itself to the use of water-based inks that do not rub off when handled. Flexography represents approximately 20 percent of all printing today, and its use is growing.

Screen-Printing

Screen-printing is the simplest of the traditional printing processes. The image to be printed is formed on a screen made of synthetic fibers over which a stencil is placed that represents the non-image areas. The area of the screen not covered by the stencil represents the image area because it is here where ink can pass through the screen and onto the substrate.

Stencils can be formed in a number of ways. One way is by photographically exposing—through negative or positive film—a light-sensitive emulsion applied to the screen. When developed, the image and non-image areas are defined. Stencils can also be formed by applying pressure-sensitive stencil material on the screen or by "painting" a liquid stencil on the screen. Once the stencil is formed, the screen is brought in contact with the substrate, ink is placed on the screen, and a squeegee drags the ink over the stencil and the entire screen. The ink that is not blocked by the stencil will go through the screen and onto the substrate to form the printed image. The process uses a porous mesh stretched tightly over a frame made of wood or metal.

Many conditions such as composition, size and form, angle, pressure, and speed of the blade (squeegee) determine the quality of the impression made. Proper tension of the frame is essential for accurate color registration. At one time, most blades were made from rubber that was prone to wear, subject to edge nicks, and tended to warp and distort. Most blades are now made from polyurethane that can produce as many as 25,000 impressions without significant degradation of the image.

A significant characteristic of screen-printing is that a greater thickness of the

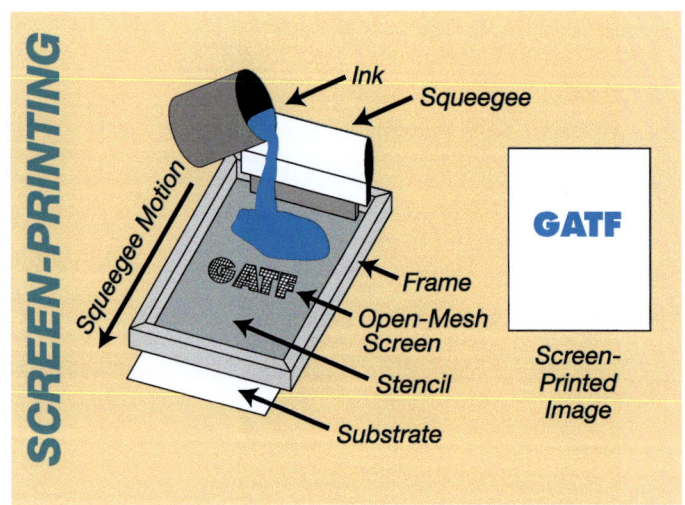

The principle of screen printing, in which ink is forced through openings in a stencil to create an image.

M&R ALPHA 8 Automatic Oval screen-printing press

ink can be applied to the substrate than is possible with other printing techniques. This allows for some effects that are not possible using other methods. Because of the simplicity of the process, a wider range of inks, including solvent-based, water-based, plastisol, and UV-curable is available for use in screen-printing.

The use of screen-printing has grown slightly in recent years because production rates have improved. This has been a result of the development of automated and rotary screen-printing presses, improved dryers, and UV curable ink. The major chemicals used include screen emulsions, inks, solvents, surfactants, caustics, and oxidizers used in screen reclamation. The inks used in this process vary dramatically in their formulations.

Following the screen-printing process, the printed product is placed on a conveyor belt to a drying oven or UV curing system. The rate of screen-printing production was once dictated by the drying rate of screen-printing inks. However, due to advances in automation, rotary-style presses, drying, and UV-curable ink technologies, the production rate has greatly increased.

There are three types of screen-printing presses: flatbed (the most common), cylinder, and rotary. Until relatively recently, all screen-printing presses were manually operated. Now, however, most commercial and industrial screen-printing is done on single-color and multicolor automated presses.

The screen process lends itself to printing that does not require long runs, and is often used for printing T-shirts and other textiles such as nylon and cotton, as well as for printing short-run posters, bumper stickers, billboards, labels, decals, and signage. It can also be used on electronic circuit boards, glass, leather, wood, and ceramic surfaces. Its advantage over other printing processes is that the press can print on substrates of many shapes, thickness, and sizes. Screen printing represents under five percent of all printing, and its use has experienced gradual growth in the recent past.

Combination Printing

Combination or hybrid printing involves printing presses that use two or more printing processes. For example, security printing—such as scratch-off lottery tickets—typically employs printing presses having three or four processes such as lithography, gravure, flexography, and inkjet printing. Package printing often involves combination presses to create different effects of solids, screens, or foil finishes.

A Stork RSI rotary screen print module in operation on a combination, or hybrid, press. (Image courtesy of Stork Prints America, Inc.)

In 1995, the DI (Direct Imaging) press combined offset printing and on-press digital imaging of the printing plate. (Image courtesy of Heidelberg.)

Combination printing is not a new phenomenon. Gutenberg combined several existing technologies with his own to invent a process that lasted for centuries.

Another popular form of combination printing involves adding inkjet printing to four-color web offset printing. This allows for the printing of high-quality publications, catalogs, and direct mail, where each individual piece can be personalized, using variable data printing or VDP.

A Kodak Prosper inkjet head, combined with high-speed web offset to imprint each page with personalized information. (Image courtesy of Eastman Kodak.)

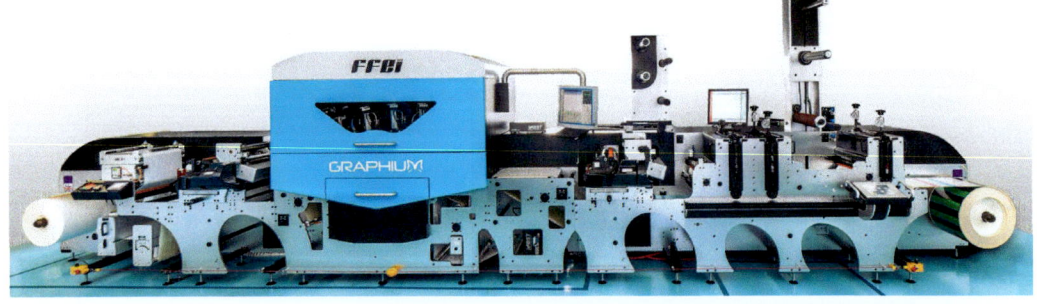

A Graphium UV press, combining flexo and digital inkjet. (Image courtesy of Fujifilm.)

Digital Printing

Today, almost all printing involves a high level of digital functionality. Conveying text and graphics to a device that can reproduce them at high volumes (a printing press) involves increasingly fewer manual steps.

Of the "conventional" printing types—offset, gravure, flexography—only the physical printing process differs from that of what we know today as digital printing. Those are based primarily on inkjet and electrophotographic or EP technology.

Most of the digital printing processes discussed in this section are "pressure-less" in nature. For example, EP printing does not use traditional wet printing ink but use toner that is directed to the image area on the substrate via electrostatic charges. Inkjet devices, as the name implies, apply their inks directly to paper from a nozzle or similar device. Hence, with inkjet and electrostatic or EP digital printing, there is no printing plate squeezing the ink onto a blanket or directly onto the substrate.

Digital Print Benefits

Besides their different imaging methods, digital printing presses are also able to more easily accept jobs directly from a computer system, and produce a printing plate or image a cylinder "on-the-fly." They can produce jobs with shorter run lengths and greater variability than conventional presses could cost-effectively handle.

Digital printing has come of age in the early 21st Century. Digital presses now produce images that rival offset

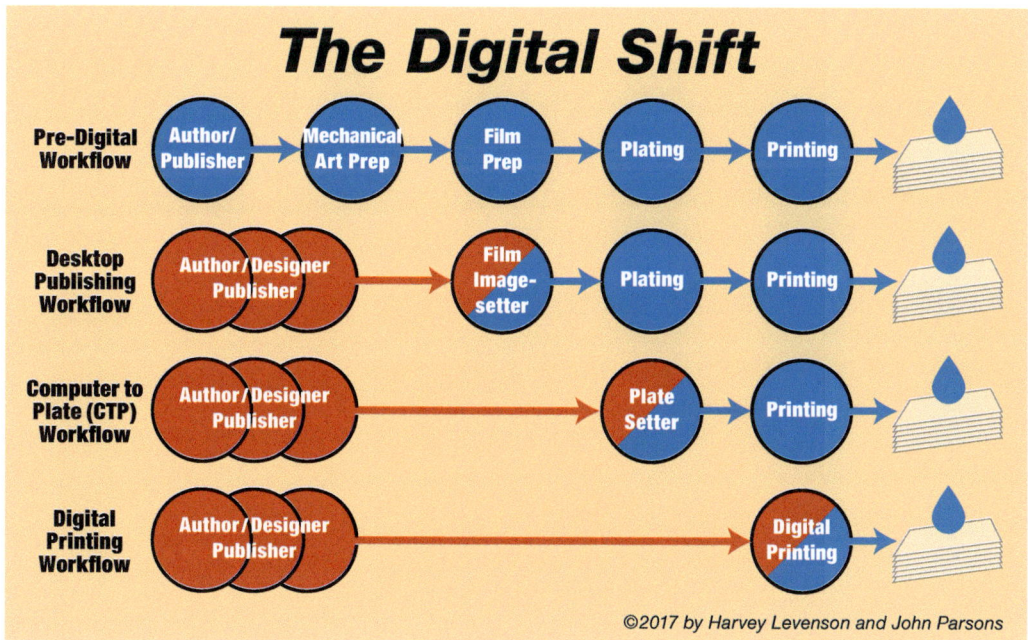

All printing, even today's inkjet and EP devices, employs a physical process to put ink or toner on a substrate. The difference is where the digital process stops and the mechanical process occurs. Before the desktop publishing phenomenon, every step was mechanical (shown in blue). With DTP, early steps in the workflow were combined digitally. Then, with each successive advance, the digital process (in red) moved closer to the actual moment an image was printed.

Introduction to Graphic Communication | 147

lithography, and lower cost-per-page pricing makes them a viable printing alternative.

Digital printing represents a growth area for many graphic communication companies. Digital printers do mostly digital printing, but some do traditional printing as well. They are also producing a growing amount of full-color printing to complement the large volume of black-and-white printing traditionally associated with digital printing through companies and corporate data centers where large volumes of transactional printing is produced, including bills, statements, policies, and reports.

> While the cost-per-page of digital color is currently too high to compete with offset for long runs, that cost will continue to fall, eroding the role of conventional printing.

In these market segments, black-and-white printing used to be predominant. Today, color is typically used in their documents. In some cases, companies print variable black text on a shell that has been preprinted in color using offset presses. Direct mail printers who previously printed almost exclusively in black when it came to digital printing have now incorporated full-color digital printing into their production mix.

The majority of digital printing companies are small, typically employing between five and ten people. They serve a variety of industries, including business and financial services, retailers, nonprofit organizations, and education and government organizations. In an industry where many firms are still fighting to survive, digital printers are thriving.

Digital printers produce a variety of printed products. Books and manuals make up the largest page volume, printed predominantly in black-and-white and sold at a relatively low cost per page. The revenue makers and growth areas are color applications, such as brochures, sales collateral, and direct-mail printing. Individual companies have also developed specialized color markets, such as business cards, short-run posters, and calendars.

One reason digital printers are growing—in an environment where other printing companies cannot—is the availability of affordable digital color. Their printed products were once impractical using traditional printing equipment. But with color digital printing costs dropping below ten cents per letter-size page, the economics have become more attractive. While the cost-per-page of digital color is currently too high to compete with offset for long runs (over 20,000 pages), that cost will continue to fall, further eroding the role of conventional printing.

The VDP Process

One unique aspect of digital printing is the ability to incorporate Variable Data Printing or VDP into a job. Because each sheet or page is imaged at the moment it is printed, it can literally be different from any other page in the press run. VDP workflows are dependent on the system's

VARIABLE DATA PRINTING (VDP)

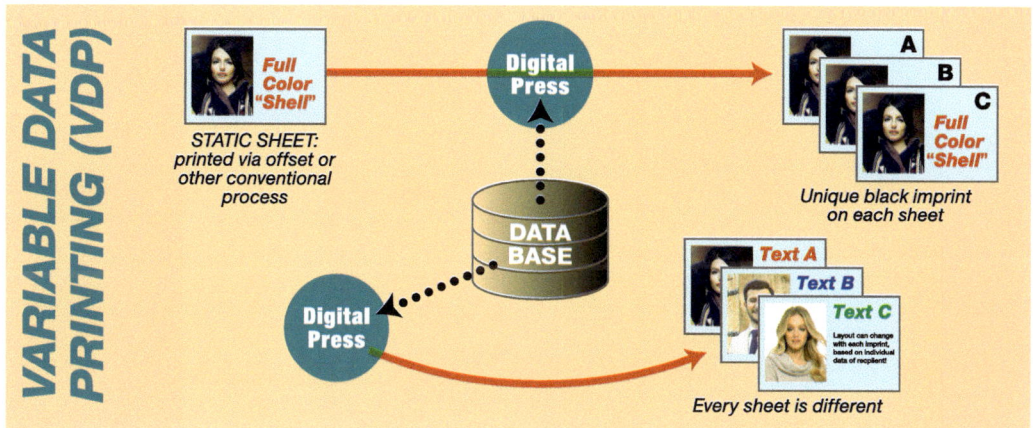

Variable Data Printing (or VDP) is based in the principle that an inkjet or EP press images each page individually from digital data. The most common technique is to imprint variable data—such as name and address information—on a pre-printed color "shell." The other, which requires greater computing power, is to create a unique page for each impression, based on unique data for text and images.

ability to handle high data volumes, and rapidly respond to each change without unduly slowing down the press. There must also be reliable and secure data handling processes in place, so one customer does not receive a printed piece clearly intended for someone else.

The benefits VDP to marketers and publishers are obvious—and only possible with digital printing. (Note that each copy of this book has a unique, data-generated serial number printed on the inside front cover.) In the past, relatively few digital printers utilized the potential of VDP fully, although the technology has been available for years. That number is growing, however.

Previously, digital printing equipment vendors promoted VDP because, in certain applications, variability justified the high cost of digital color pages. For static pages, digital printing was just not able to compete with offset lithography, except for extremely short runs. Today, however, VDP is only one aspect of digital printing, which is continuously displacing conventional processes in most printing applications.

Digital Printing Engines

There are different printing engines that drive digital presses. Digital systems consist of inkjet, electronic, electrophotographic (EP), magnetographic, ion deposition, light-emitting diode (LED), liquid crystal shutter (LCS), electron beam imaging (EBI), thermal, and electrostatic printing. These are all processes—used mainly for short runs and printing variable or personalized information—in which data representing the images are in digital form until the moment of actual imaging.

Inkjet printing. In the early days of the technology, inkjet printing was used mainly for simple, monochrome printing such as variable addressing, barcoding, computer letters, sweepstakes forms, and personalized direct mail advertising. Increasingly, however, inkjet devices are increasing in speed, versatility, and cost-effectiveness, putting them in contention for an increased share of the color printing market.

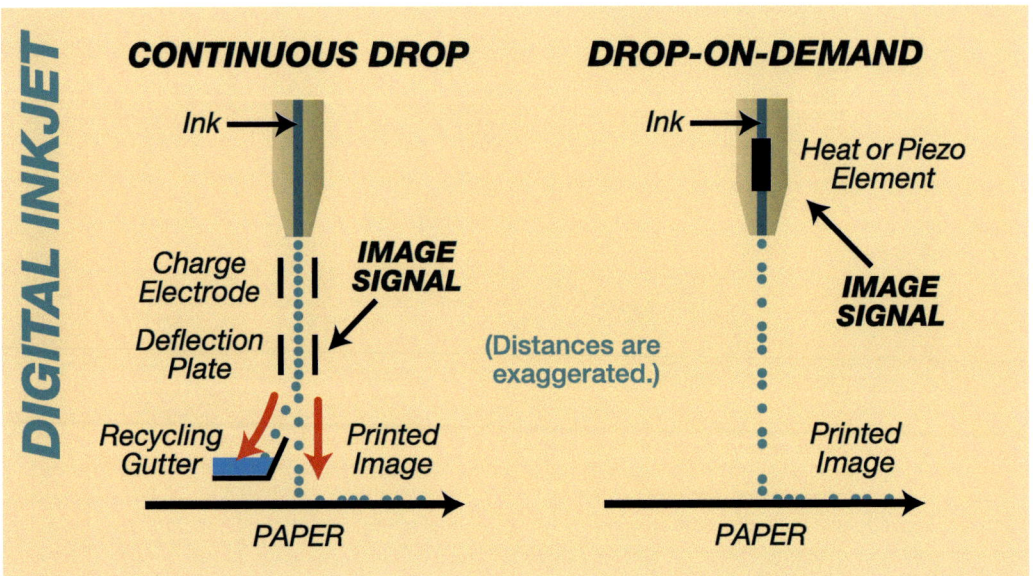

Inkjet printing uses either a continuous drop process (where an electrical charge deflects droplets that are not to be imaged) or drop-on-demand (where an electrical charge initiates the correct droplets).

During the inkjet printing process, microscopic droplets of ink are squirted onto a substrate from a print head containing one or more nozzles. One type of inkjet printing is the **continuous drop** process, where a stream of ink droplets is forced through a nozzle under constant pressure. The ink droplets are deflected to the image area via electrostatic charges. Unneeded droplets are not charged and are deflected into a gutter for recycling

One disadvantage of inkjet printing is that the inks are often water-soluble and can easily smudge when subjected to moisture (applying a protective coating can prevent smudging).

The other major inkjet printing technology is the **drop-on-demand** process, where ink drops are forced through the nozzle only when needed. There are three types of this process. One is **piezoelectric**, where an oscillating crystal produces an electric charge that causes the ink drop to be expelled. Another, **bubble jet/thermal liquid ink,** where an electric charge is applied to a small resister causing a minute quantity of ink to boil and form a bubble that expands and forces the ink droplet out of the nozzle; The third, **solid ink**, involves a wax-based ink that melts quickly and solidifies on contact with a substrate.

Electrostatic and Electrophotographic (EP) printing. Electrostatic and electrographic (EP) printing are similar in that they relate to the principles of xerography, where toner particles are used to form an image. In electrostatic printing, unlike EP printing, there is no print drum. Toner particles are attracted directly to the paper through controlled conductivity. No optical system is used. The copier glass is exposed at once and an electrostatic charge is directly deposited onto the paper. The toner is fused to the paper through hot air. The electrostatic process is typically found in office printers.

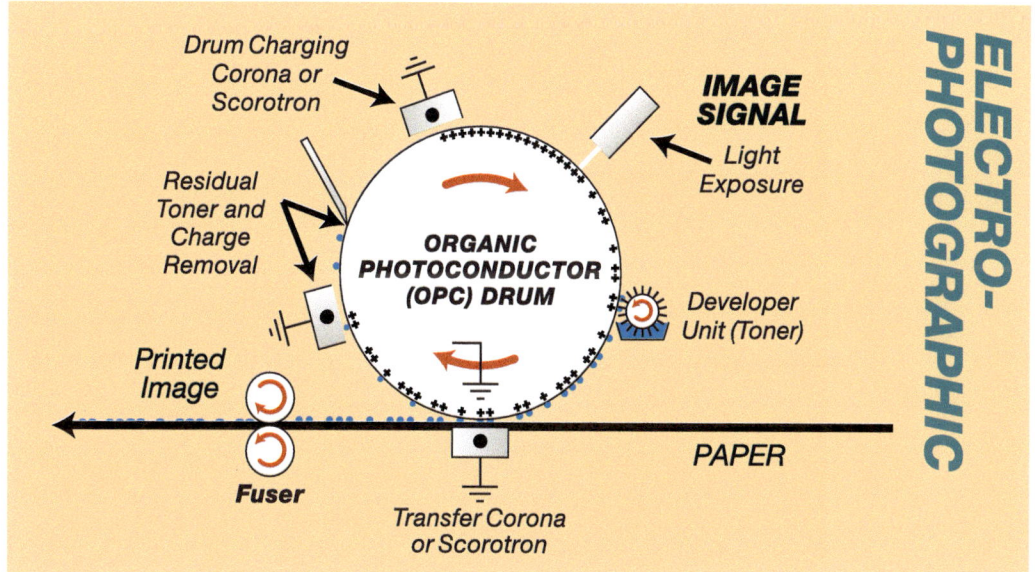

The operating principle of an electrophotographic digital press like that of an office copier. Charged toner particles adhere to selectively charged portions of a drum, and are then transferred to the paper substrate.

Electrophotographic or EP printing uses a print drum and photoconductor that is charged by a corona discharge and then imaged by a moving laser light beam modulated by digital signals from a PostScript-based or PDF-based digital imaging system. In this process, a laser beam is focused on a rotating mirror that deflects the beam through a focusing lens that forms a latent image on a photoconductor. EP devices are faster that electrostatic printers and used for both monochromatic and full-color, production printing.

Other Digital Print Processes. There are a host of secondary technologies that relate to the print engine driving the digital printing processes. These include electron beam imaging (EBI), ion deposition, light-emitting diode (LED), liquid crystal shutter (LCS), magnetographics, and thermal printing. Typically, they are "pressureless" in nature, are components of print heads, have few moving parts, and emit high-intensity lights and electrical signals to convert data into printed pages. Unlike inkjet and EP technologies, these processes are largely confined to office printers, and are not widely used in high-production environments.

Wide-Format Digital Printers

Wide-format digital printers provide individuals and companies with access to inexpensive large-sized prints. A growing number of companies manufacture systems that produce full-color digital prints ranging in size from 36 in. to 54 in. However, systems that produce larger sizes are also available. There are two components to these systems: a large-format printer and a RIP. Large-format digital printing systems print on a variety of substrates including paper and Mylar and use engines represented by inkjet, electrostatic, or thermal wax transfer technology.

Large format devices typically use a drop-on-demand inkjet process. Applications include contract color proofing (see Chapter 6) and a variety of art reproduction and signage uses, from poster-sized sheets to large "wraps" for vehicles and even buildings. The inks can be aqueous or solvent-based. Traditional wide-format inks require the use of specially-coated papers. More recently, however, Ultraviolet- or UV-curable inks are used in order to print on uncoated substrates (among other benefits). Water-based latex inks—introduced by HP—are also used in wide format printing, as an environmentally friendly way to produce outdoor media.

This market is growing rapidly and finding its niche in the "quick printing" and on-demand printing industry segments. Users of this technology also include advertising agencies, screen printers, and in-plant printing departments of many businesses. The applications are numerous and include items such as art-on-demand, backlit signage, transportation advertising displays, engineering drawings, maps, murals, posters, window graphics, and more.

Digital Printing Presses

Since the introduction of the first models in the early-to-mid 1990s, the evolution of digital color printing presses has accelerated beyond the imagination of color theorists and technicians. The paradigm that printing technology depreciates over a ten-year period has been replaced with a "scientific revolution" reducing the cycle of technological transition in the graphic arts from a decade or more to a few years or even months.

The technology of computer-to-film, bypassing the laborious tasks of graphic arts photography and image assembly, was a monumental step toward automating the printing processes in the late 1980s and early 1990s. Computer-to-plate technology, and its popularity in the 1990s, was an extension of this technology. However, computer-to-press technology has provided complete integrated printing systems putting the author, copywriter, and artist in the position of producing finished printed products.

The Xerox DocuTech represented the first wave of such technology in the production of black-and-white printing. However, soon after DocuTech's introduction in the late 1980s, other

ValueJet 1204, a wide-format digital printer. (Courtesy Mutoh America, Inc.)

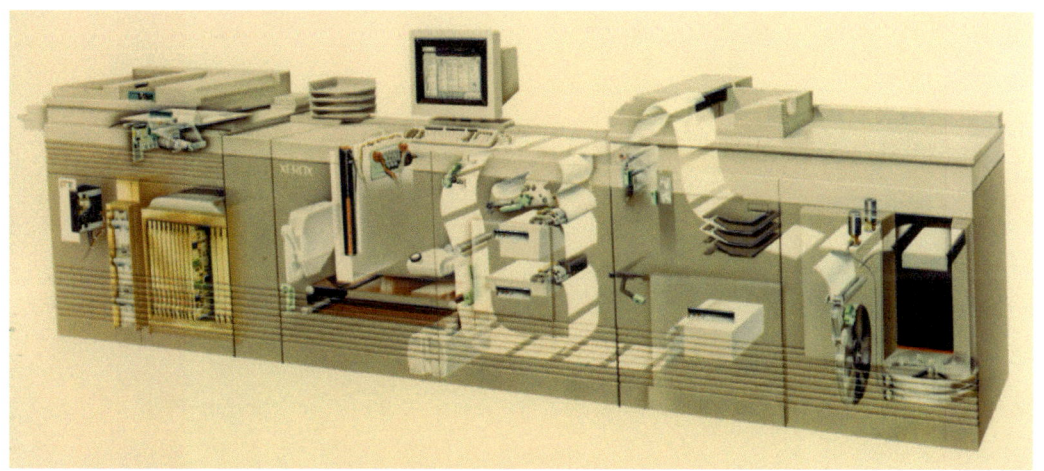
The Xerox DocuTech 135, introduced in 1990 (Image courtesy of Xerox Corporation.)

manufacturers saw the future of direct-to-press technology in color markets and nearly all systems afterwards addressed this demand. Heidelberg, the first to introduce a direct-to-press color system in its GTO-DI, was quickly followed by Indigo, Xeikon, Agfa, and others. By the year 2000, Heidelberg had already introduced several generations of direct-to-press systems. Today, a growing number of companies are producing digital color printing systems to compete with commercial printing presses.

Much of the accelerated development in this area is the result of mergers, joint ventures, and acquisitions where companies acquire existing technology rather than reinventing it. Thus, two or more technologies are brought together to create improved systems, speeds, and technological advances.

Digital printing has resulted in a wide range of new companies serving the industry's equipment needs. While traditional companies such as Heidelberg and manroland have entered the digital arena and then left it to focus on their core traditional technologies of offset lithographic printing presses, companies including Canon/Océ, EFI, Fujifilm, Hewlett Packard, KBA, Kodak, Komori, Konica Minolta, Landa, Memjet, Pitney Bowes, Ricoh, RISO, Scitex, Screen Americas, Xanté, Xeikon, Xerox, and others all are now manufacturers of printing technology, but of the digital variety. These companies provide not only hardware but also intangibles such as software and digital front-end workflow systems. Digital workflow strategy and production workflow processes are as important as hardware in driving printer production and productivity.

On the hardware side, systems that handle a greater variety of substrates, including very lightweight paper, have been developed. The concept of "universal copier/printer" devices has been developed. These are devices that output color and monochrome pages at competitive costs with dedicated color and monochrome printers. Adding value by integrating all services from front-end to printing to finishing is key to digital printing, as is flexibility and process improvement.

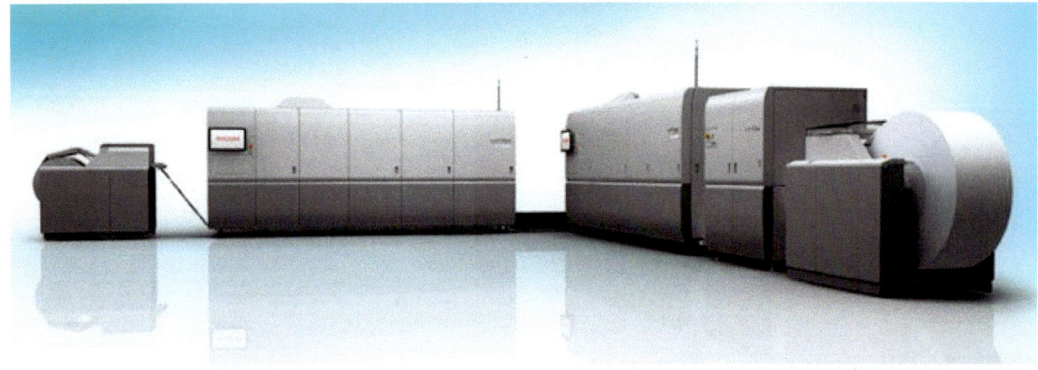

Ricoh Pro VC60000 inkjet press, the device used to print this book. (Image courtesy of Ricoh USA Inc.)

A modern printing operation today also uses the Internet and World Wide Web as common business tools for receiving information and for executing related business functions. Innovations in RIP, server, and workflow technology are making it easier to integrate digital devices into almost any printer's array of output services.

Digital printing has caught up with offset lithography in quality of output, and the digitally driven convergence of the two output technologies is already underway. Within the range of applications that they were designed for, digital presses compare with offset in color in such a way that even trained eyes can no longer distinguish the differences. Digital output systems now complement conventional offset lithographic equipment.

Digital Device Comparisons

Digital color printers and copiers. Color copiers represent the broadest range of manufacturers and features. Once relegated to the office environment, new digital color printers have become ubiquitous in printing companies as well. They employ basic laser imaging technology to charge an image on either drums or belts, from which the developed CMYK image is ultimately transferred to paper. Any project that demands process, full-bleed color in runs of up to 5,000 can be imaged on these high-resolution, toner-based devices.

Digital color production presses. The Kodak NexPress and Xerox DocuColor iGen series represented a breakthrough for commercial printers when these technologies were first introduced over a decade ago. They brought short run/JIT, fast turnaround, web enablement, and

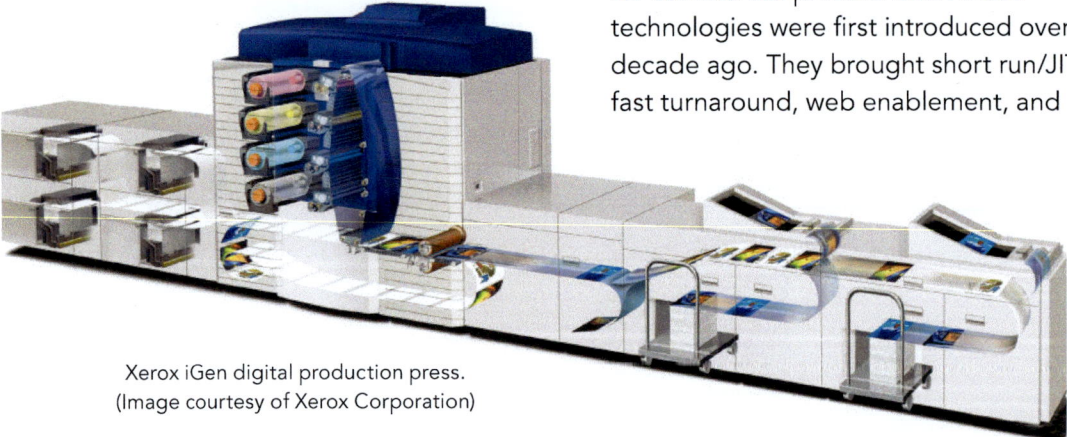

Xerox iGen digital production press. (Image courtesy of Xerox Corporation)

personalized printing into the press department. Although there are individual differences, these toner-based presses are fast and feature extremely high print quality and the ability to deliver collated sheets at the end of the press. They run smooth and textured papers in a range of sheet sizes and basis weights, with the ability to mix stocks in a single run.

Xeikon-engined printers. A toner-based web press, the Xeikon-engined printer feeds its web through a series of drums, each charged with the image and each applying one process color. Process color toner is fused to the sheet with adjustable heat and pressure; changing the heat and pressure levels results in more or less gloss in the toner.

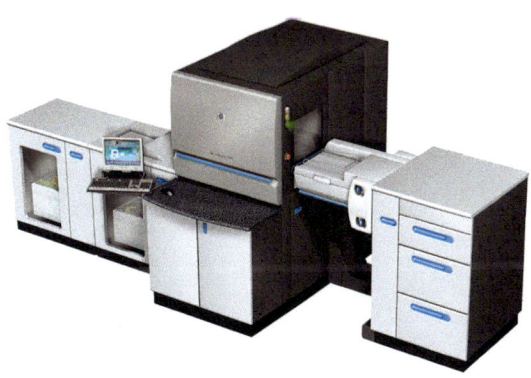

HP Indigo 5000 sheetfed press
(Image courtesy of Hewlett-Packard)

HP Indigo presses. These presses use one imaging drum and patented liquid ElectroInk. Both drum and ink are charged; ElectroInk adheres to the image area on the drum and a blanket transfers the image to paper. No ink is left on the blanket. The plate charge is cleared and the process is repeated for subsequent colors. This process supports fully variable data and very high-resolution images. The HP Indigo press can print on a wide range of substrates, but for optimum performance pretreated paper often improves toner adhesion.

DI presses. These sheetfed presses were popular for a short time but were replaced by presses with a more fully digital workflow. The DI presses worked like offset presses with an electronic twist: directed by digital data, pre-mounted plates were imaged with a laser right on the press, reducing makeready time to minutes. They were ideal for process color jobs from 500 to 10,000, and produced high-resolution offset images using

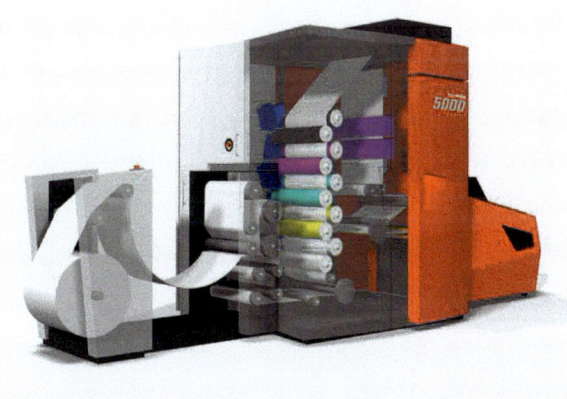

Xeikon 5000 web-fed digital press (Courtesy Punch Graphix)

An on-line cutter trims pages to length. For optimum performance paper must be scripted in order to establish set points on the equipment for heat and pressure as well as other key characteristics. This high-resolution duplexing printer has full variable-data capability—meaning that some or all of the text or images can be changed from one document to the next.

Introduction to Graphic Communication | 155

Landa W10P Nanographic Printing Press (Courtesy Landa Corporation)

traditional inks and were ideal for promotional and sales literature, publications, and even packaging.

High-speed and high-volume digital presses are used for documents requiring hundreds of thousands or even millions of copies. They have variable-data capabilities and are very popular for transactional documents such as telephone bill, cable television bills, utility bills, and much more. They often have the capabilities of producing full-color images along with personalized messages directed specifically to the recipient.

Nanography. One of the latest of the digital printing processes is Nanography, developed by the Landa Corporation. It has enough attributes that differ from other digital printing processes that makes it worthy of notice. The process uses pigment particles under 100 nanometers in size. (A nanometer one-billionth of a meter. This produces images with ultra-sharp dots of high uniformity and high gloss. The process begins with the jetting of billions of droplets, not ejected directly onto the substrate—as in the typical inkjet process—but onto a blanket from ink ejectors positioned one to two millimeters away.

Printing Economics

Several trillion pages are printed in North America each year, mostly on offset equipment. Much of that volume will eventually be a candidate for digital printing. For at least half of those pages, printing cost is the primary obstacle to shifting from offset to digital. And the cost of digital printing is finally reaching a level at which it can compete for a growing number of those pages. While this represents a growth opportunity for digital printing, it still requires a good marketing strategy, an efficient workflow, and good customer service.

Besides the improvements in color quality and VDP, digital print offers other important economic incentives. Among these are the "on-demand" nature of the technology. Because each job can be economically produced in much smaller quantities, there is less need to stock inventories of printer materials—or dispose of outdated materials. This allows for "Just In Time" printing and fulfillment, and resulting cost savings. The actual equipment and labor costs for digital printing are also significantly lower than they are for conventional processes.

Perhaps one of the most common arguments for the use of digital printing is the reduction in "makeready" overhead.

Conventional presses often require the operator to print (and then discard) multiple copies of a job before it is acceptable. This is the result of adjusting ink levels and other variables, measuring the results on a press sheet, and repeating the process until the desired results are achieved. Digital presses, in theory, require far less makeready, and are, therefore, more economical.

There are a growing number of digital printing presses affecting print production and distribution. Such systems currently appeal to niche markets such as on-demand, variable-data, and short-run "quick" printing services. These offer both black-and-white and color printing of commercially acceptable quality on a relatively small range of page sizes.

However, this trend is increasingly true for more traditional commercial printing companies presently dominated by offset and other conventional processes. This promise will be realized as digital printing systems continue to improve in the quality of the printed image, the size of sheet or web accommodated, and the speed of production. For all these reasons, digital printing represents the fastest growing printing process.

The Future of Printing

Offset, gravure, and flexography are likely to endure for a long time, but the incursions of digital printing will continue to dominate the graphic communication industry. A sampling of Original Equipment Manufacturers (OEMs) at major printing events illustrates this. At DRUPA 2000 and PRINT 01, exhibition halls were filled with conventional printing equipment, and relatively few digital devices. Today, the situation is reversed. Most of the traditional printing OEMs and many new players are promoting digital print as the future of the industry.

The future of digital printing will undoubtedly include significant improvements and new developments in web-to-print, print-to-web, printed electronics, and even more exotic applications like 3D, inkless, and water transfer printing.

The book you are presently reading is such an application. It was printed on a digital press, using many of the technologies described in this chapter. It is also used to trigger a variety of non-print content and interaction.

As in Gutenberg's day, the printing press is still the central "engine" of graphic communication. While the choice of processes has never been greater, the end results—mass-produced text and images on a durable, practical surface—remain the same.

9
Postpress and Finishing

Postpress and Finishing

Chapter Preview

Cutting, Folding, and Assembly

Binding Methods and Covers

Finishing Techniques and Special Effects

Automation and the Digital Bindery

Interactive Media: www.igcvideo/Chapte9   Web Links: www.igcbook/Chapter9

Overview

Once a piece is printed, it must be made suitable for ite indended use, in the demanding process known as postpress. Options for doing all this are as varied as there are uses for printed material—from the simplest flyer or mailer to a richly adorned book cover or a luxury product package.

Postpress and finishing are often considered the most important part of the printing process. If all else is done well, a postpress error will ruin the entire job.

Postpress and finishing began well before Gutenberg's time. Early printed books and even handwritten manuscripts had to be finished for optimal use. A multi-page book had to be sewn, glued, and bound together with a protective, frequently decorated cover, in order to keep the valuable content secure.

As printing became a manufacturing process, postpress technology advanced accordingly. This chapter outlines the basic processes used today. In order, these are cutting, folding, assembly, binding, and finishing. Many of these steps are mandatory, while some are optional.

The chapter also covers cost-saving advances made possible by digital workflows and automation. As with other facets of graphic communication, postpress is evolving from a traditional craft or trade business model to one of automated manufacturing.

Cutting, folding, and binding are only part of the postpress process. Other finishing options can make ordinary printing look extraordinary. (Image courtesy of MGI Digital Technology.)

Introduction to Graphic Communication | 161

Postpress Basics

With very few exceptions, printing press output is not ready for use. A single press sheet typically contains multiple pages, in seemingly random order. This output is easy to stack and ship but clearly not user-friendly. Postpress and finishing include all the processes that turn sheets of printed material into the books, brochures, and boxes we use every day.

The processes described in this chapter can be either offline or inline. The former simply means that stacks of printed sheets are physically moved from one device (the press) to the next logical cutting, folding, binding, or finishing machine. As the name suggests, inline postpress involves connecting the right machines—using a conveyor belt or other mechanism to transport sheets rapidly through successive steps in the process.

Postpress consists of four major processes: cutting, folding, assembling, and binding. All four are not always necessary. For example, simple folded brochures do not undergo binding. However, there are also other steps including, but not limited to, die-cutting, embossing, debossing, foil stamping, and scoring.

Cutting

All paper used in printing is manufactured in large rolls, and sometimes cut into press-ready sheets, as discussed in Chapter 7. This means that print output must be cut or trimmed to its final size.

Although cutting is generally considered a postpress operation, most lithographic

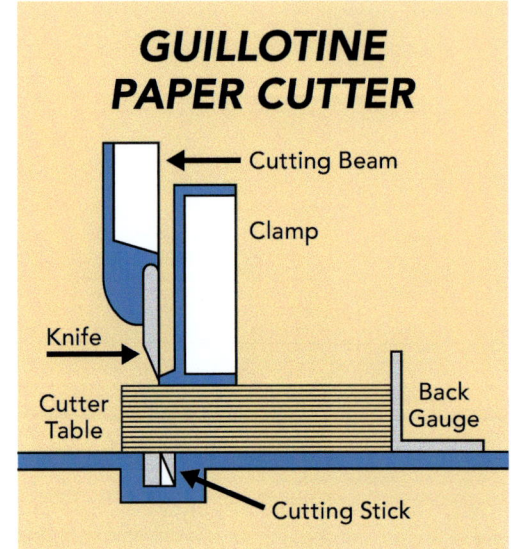

and gravure web presses have integrated, inline cutting mechanisms, as well as equipment to perform related operations such as scoring and perforation.

A machine used for cutting press sheets into individual pages—typically offline—is called a guillotine cutter. Built in many sizes, capacities, and configurations, these machines consist of a flat bed or table that holds the stack of paper to be cut. At the rear of the cutter, the stack rests against an adjustable back guide, allowing the operator to accurately position the paper. Side guides or walls of the cutter are at exact right angles to the bed. A clamp is lowered into contact with the top of the paper stack to hold it in place while it is cut. An electrically-powered hydraulic pump drives the cutting blade itself.

In recent applications of the Job Definition Format (JDF), covered later in the book, guillotine cutters can be programmed to know the cutting specifications of the next job coming from a printing press, automatically presetting the cutter to expedite the process.

To assist the operator in handling large reams of paper—which can weigh as much as 200 pounds—some tables blow air through small openings in the bed of the table. This lifts the stack slightly, providing a near frictionless surface on which to move the stack.

The cutter operator uses a cutting layout to guide the process. Typically, the layout is one sheet from the printing job that has been ruled to show the location and order of the cuts to be made.

For many types of high-volume printed work, particularly books and publications, cutting is a highly-automated process. An inline, three-knife trimmer, similar in principle to the guillotine cutter, is often used *after* the binding process described later in the chapter.

Folding

Except for mailers, posters, and other single-page jobs, printed sheets must be folded into the correct sequential order for reading. Each press sheet is folded to create a *signature*—a term originating from medieval bookbinding. The pages in a signature are arranged and numbered during the prepress phase (see Chapter 5) so that they will appear in the correct order after the folding process.

Signatures for books and publications always contain an even number of pages, based on the size of the sheet and the number of folds. The smallest folded signature contains four pages—two per side. Signatures of 8, 16, and even 32 pages are also used, sometimes in combination. These are gathered or *collated*—often by high-speed equipment—before the binding process.

Publication signatures are only one style of paper folding used in postpress. In fact, they are the simplest kind of folding, namely the half-fold and quarter-fold pictured here. There are innumerable

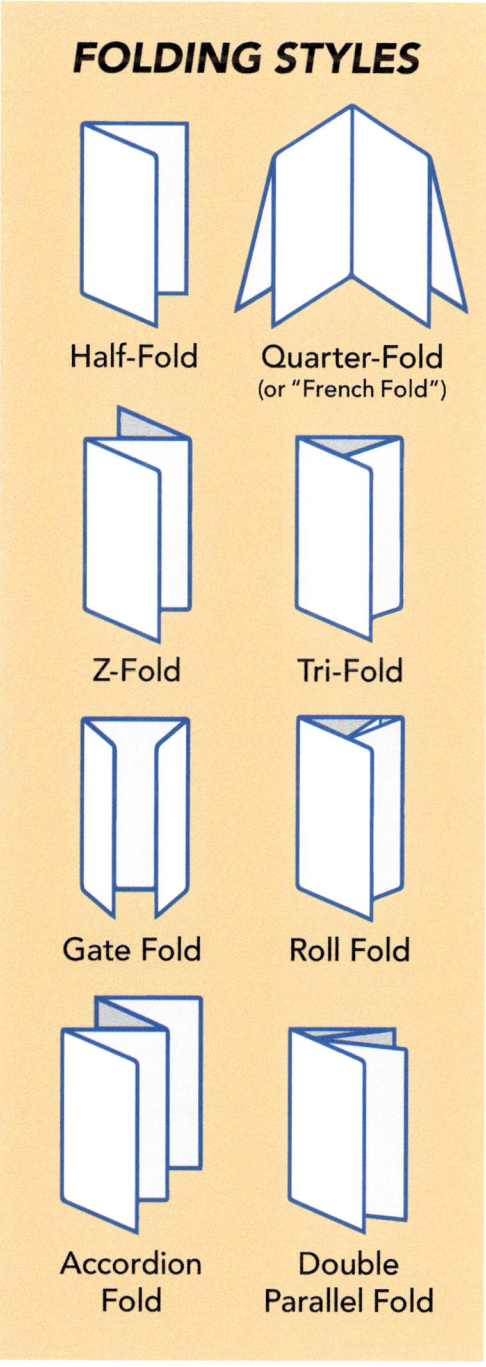

Only some of the many different styles of non-packaging paper folding.

Introduction to Graphic Communication | 163

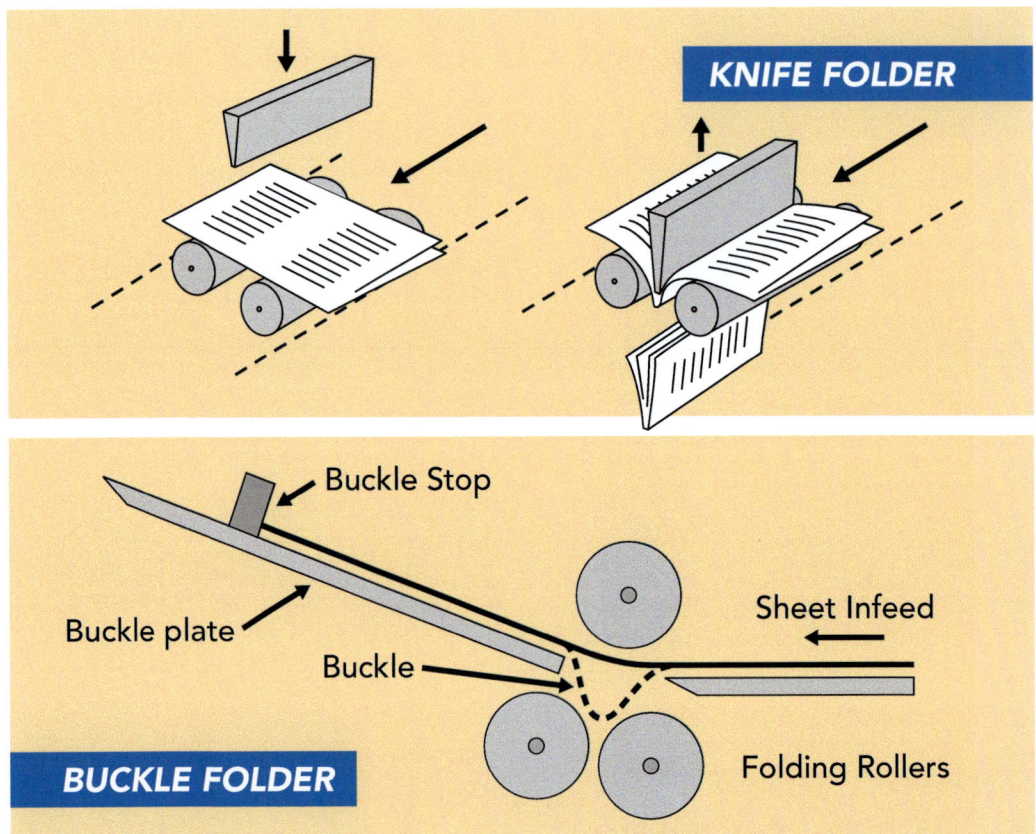

other ways to fold a printed sheet as part of the overall design.

Folding for packaging applications (see Chapter 10) add yet another level of complexity, since the resulting product is three-dimensional, and must account for gluing tabs and other necessities.

Although folding is generally considered a postpress operation, most lithographic and gravure web presses are equipped with folders. Many digital presses also employ inline folding capabilities.

There are three different types of folders used in modern printing companies. Bone folders (not pictured) have been used for centuries. These folders are simple, shaped pieces of bone or plastic that are passed over the fold to form a sharp crease. They are still used today, but only for small, high-quality jobs.

Knife folders use a thin blade to force the paper between two counter-rotating rollers. This folds the paper at the point where the knife contacts it. A fold gauge and a moveable side bar are used to position the paper in the machine before the knife forces the paper between the rollers. The rollers have knurled surfaces that grip the paper and crease it. The paper then passes out of the folder and into a gathering station. Paper paths, knives, and roller sets are stacked to create several folds on the same sheet as it passes from one folding station to another.

Buckle folders differ from knife folders in that the sheet is made to buckle and pass between the two rotating rollers of its own accord. In a buckle folder, drive rollers cause the sheet to pass between a set of closely spaced folding plates. When the sheet comes in contact with the sheet gauge or "buckle stop," the drive rollers continue to push the paper, causing it to buckle over and then pass between the folding rollers. Adjusting the position of the stop determines the distance between the fold and the lead edge of the sheet.

Assembly

The assembly process brings all of the printed and non-printed elements of the final product together prior to binding. Assembly usually includes three steps: gathering, collating, and inserting.

Gathering is the process of placing signatures next to one another. Typically, gathering is used for assembling books that have page thicknesses of at least three-eighths of an inch. Collating is the process of gathering together individual sheets of paper instead of signatures. Inserting is the process of combining signatures by placing or inserting one inside another. Inserting is normally used for pieces having a final thickness of less than one-half inch.

Assembly processes can be manual, semiautomatic, or fully automatic. In manual assembly operations, workers hand-assemble pieces from stacks of sheets or signatures laid out on tables.

Sheets or signatures are picked up from the stacks in the correct order and either gathered, collated, or inserted to form bindery units. Some printers use circular, revolving tables to assist in this process. However, due to the high cost of labor, this is limited to small jobs.

With semi-automatic assembly, stacks of sheets or signatures must be manually loaded into feeder units. Operators at each feeder station open the signatures and place them at the "saddlebar" on a moving conveyor. The number of signatures in the completed publication determines the number of stations on the machine. Completed units are removed at the end of the conveyor and passed on to the bindery.

With automatic assemblers, the sheets or signatures travel directly to the bindery machines on conveyor belts, without human intervention. These are typically part of an inline setup and are used in high-volume operations.

A Bolero B8-1 perfect binder. (Courtesy of Muller Martini)

Introduction to Graphic Communication | 165

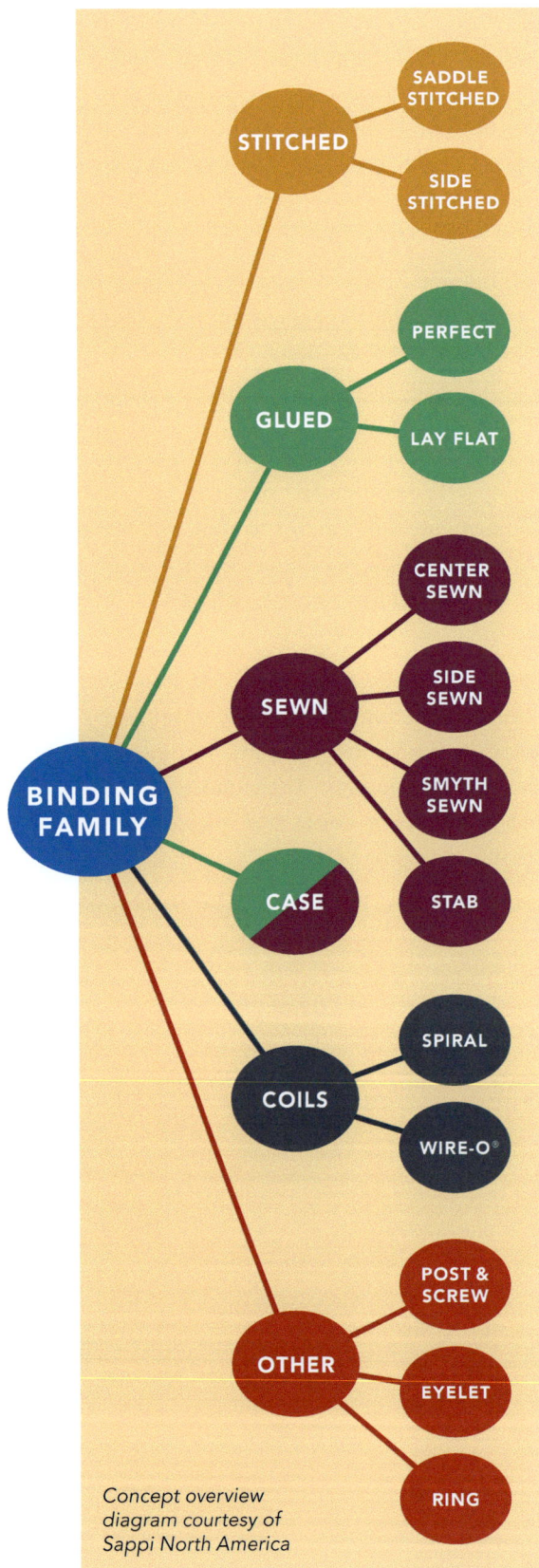

Concept overview diagram courtesy of Sappi North America

Binding Methods

Binding is the process of attaching multiple units of printed material as a single piece, either with a protective and often decorative cover, or with the sheets or signatures alone. The latter is known as "self-cover," and is typical for brochures and collateral, while the latter is common for books and periodicals.

Binding is classified according to the method of attaching pages or signatures together. The three most common types are stitched (using wire staples) glued and sewn. Case binding, a combination of gluing and sewing, is a specialized process used mainly for hardcover books. Other methods, using wire or plastic coils and other mechanisms, are typically used for business documents and certain types of publications.

In saddle stitching, one or more signatures are fastened along their folded edge with wire, staple-like stitches. The term saddle stitching comes from an open signature's resemblance to an inverted riding saddle. It is used extensively for magazines and booklets with comparatively fewer pages. Most saddle stitching is performed inline, on highly-automated finishing lines. Large, manually operated staplers are used for small printing jobs.

Adhesive binding, also known as padding, is a simple form of binding that includes detachable sheet notepads and paperback books, using a process called perfect binding. The latter must be strong enough to prevent pages from pulling out during normal use and uses a hotmelt glue with much greater adhesive strength than the water-soluble latex

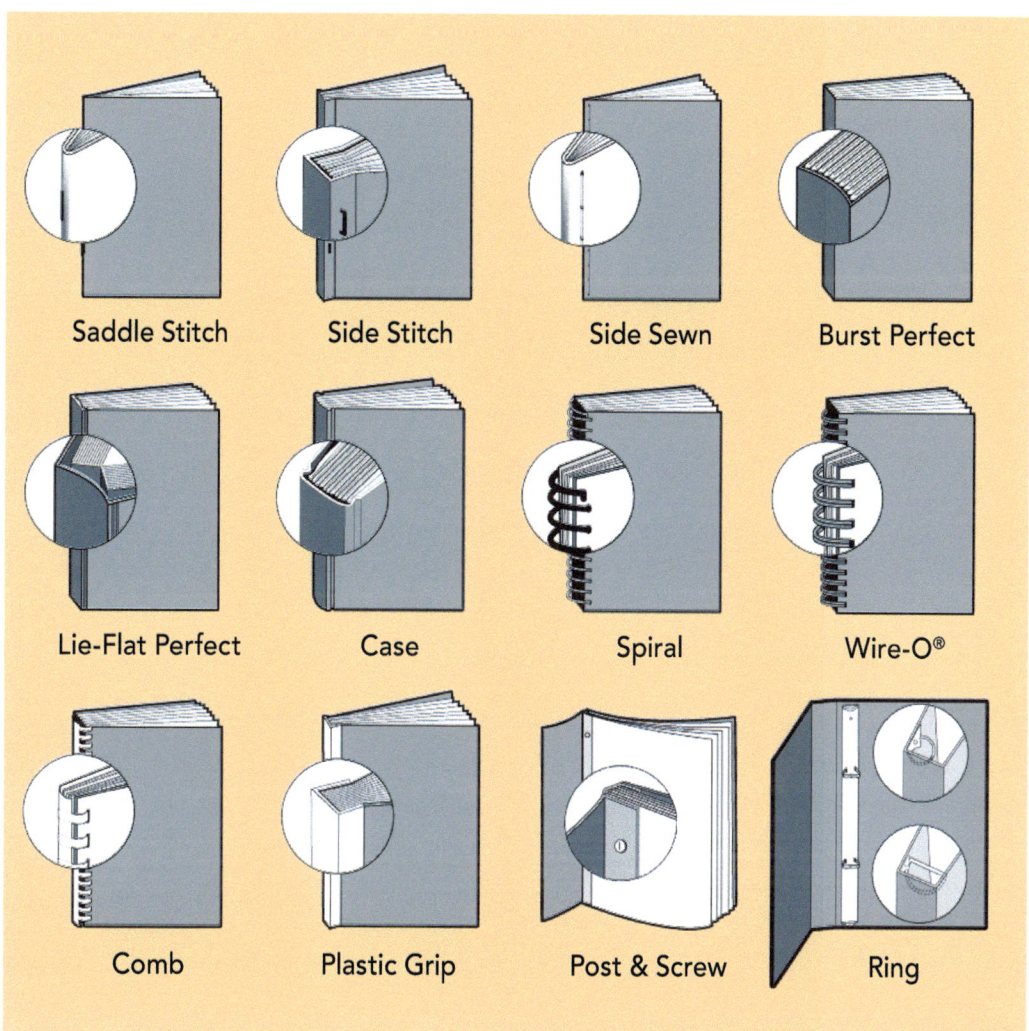

A few of the many available binding styles available in postpress.

used for notepads. In perfect binding, a piece of gauze-like material is applied along with the glue to provide added strength.

Side-sewn binding involves drilling an odd number of holes in the binding edge of the unit and then clamping the unit to prevent it from moving. A needle and thread are then passed through each hole proceeding from one end of the book to the other and then back again to the beginning point. The disadvantage of this type of binding is that the book will not lie flat when opened.

An alternative saddle binding process, Smythe sewing, is considered to be the highest quality fastening method used today. The center-sewing approach produces a book that will lie almost flat.

A common form of mechanical binding is the metal or plastic spiral technique. In this method of binding, a series of holes are punched or drilled through the pages and cover and then a wire is run through the holes. Mechanical binding is generally considered as permanent; however, plastic spiral bindings are available that

Self-covers are made from the same paper substrate as the body of the printed product. This is common for saddle-stitched booklets, flyers, and other business products. Newspapers, although unbound, are another example of a self-cover product.

can be removed without either tearing the pages or destroying the binding material. Mechanical binding typically requires some manual labor, although automated, inline processes are available.

Loose-leaf or ring bindings allow for the removal and addition of pages. The post, screw, and eyelet styles provide similar capabilities, although not as conveniently as the familiar three-ring binder.

Case-bound covers are the rigid covers generally associated with high-quality, hardcover books. This method of covering is considerably more complicated and expensive than others. Although the process is highly automated today, it is based on centuries-old techniques.

For case binding, signatures are trimmed to produce a single stack known as the book block. This includes a rounded front

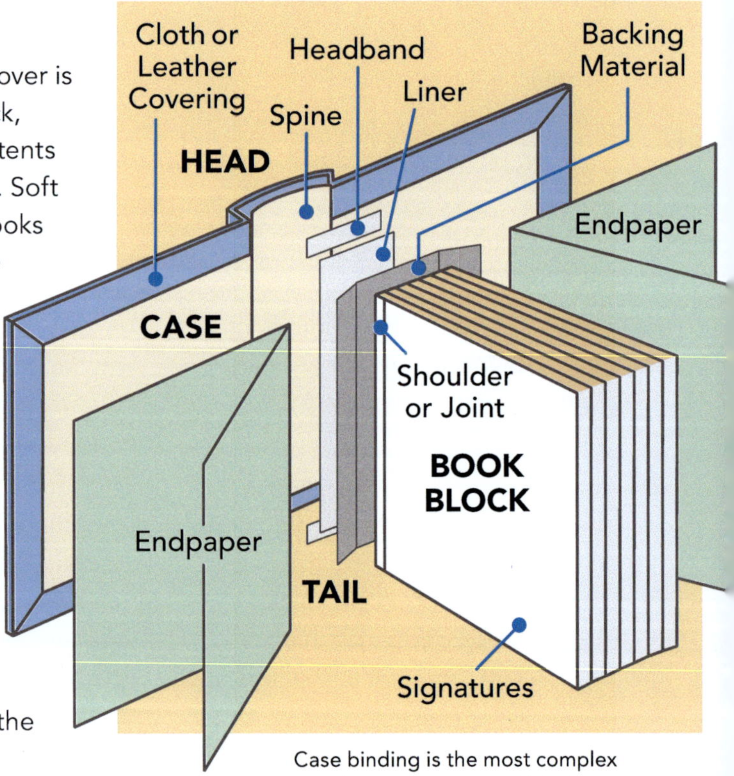

Covers

For many binding types, a cover is simply a heavier weight stock, intended to protect the contents from ordinary wear and tear. Soft covers for perfect-bound books are usually cut flush with the inside pages and attached to the signatures with glue, although they can also be sewn in place. Cover stock is typically coated on one side, to facilitate high-quality printing.

Book cover art also serves a marketing purpose, to attract potential buyers and help them feel good about the purchase. They can also be embellished with special finishing effects, such as embossing and foil stamping.

Case binding is the most complex and expensive style of binding.

(open) edge to give the finished book an attractive appearance and a back-edge shape compatible with the shape of the cover. A backing is applied by clamping the book block in place and splaying or mushrooming out the fastened edges of the signatures. This makes the rounding operation permanent and produces a ridge for the case-bound cover.

Gauze and strips of paper are glued to the back edge of the book block. These are eventually glued to the case for improved strength and stability. Bands are applied to the head and tail of the book for decorative purposes. The case is made of two pieces of thick, "binder's board" glued to the covering cloth or leather. The covering material can be printed either before or after gluing by hot-stamping or screen-printing. Finally, durable, often decorated end sheets are applied, to attach the case to the body of the book.

Finishing and Special Effects

There are still many steps to perform before a printed piece is ready for use. These range from the mundane, such as envelope insertion, to the spectacular. The latter include die-cutting (or laser cutting), embossing, debossing, metallic foil stamping, special coatings, raised surface effects, and even insertion of electronic tags or Radio Frequency Identification (RFID) chips increasingly used in packaging and other applications.

Finishing choices are determined by the nature of the piece. For example, direct mail involves only a few, highly-automatable steps, while high-end magazines or books require far more.

Inline Finishing

Historically, many finishing operations were manual, labor-intensive operations, handled within a printing company or by a trade bindery, including the following:

- **Basic bindery,** such as cutting, perforating, folding, trimming, and stitching
- **Inserting** a single piece, or multiple pieces, into an envelope or other pre-printed enclosure
- **Labeling**, either by imprinting the original piece or applying an adhesive, printed label
- **Poly-bagging** each piece in a weatherproof, mail-ready enclosure
- **Boxing or palletizing** the printed content for mailing or shipment

Even when performed within a printing company, these finishing operations generally were not integrated with the presses or with each other. Today, web-fed presses—conventional and digital—are often linked directly to inline finishing equipment, including database-driven inkjet label printers.

One of the most important results of computer-enabled inline finishing is the introduction of demographic binding, the selective assembly of a publication based

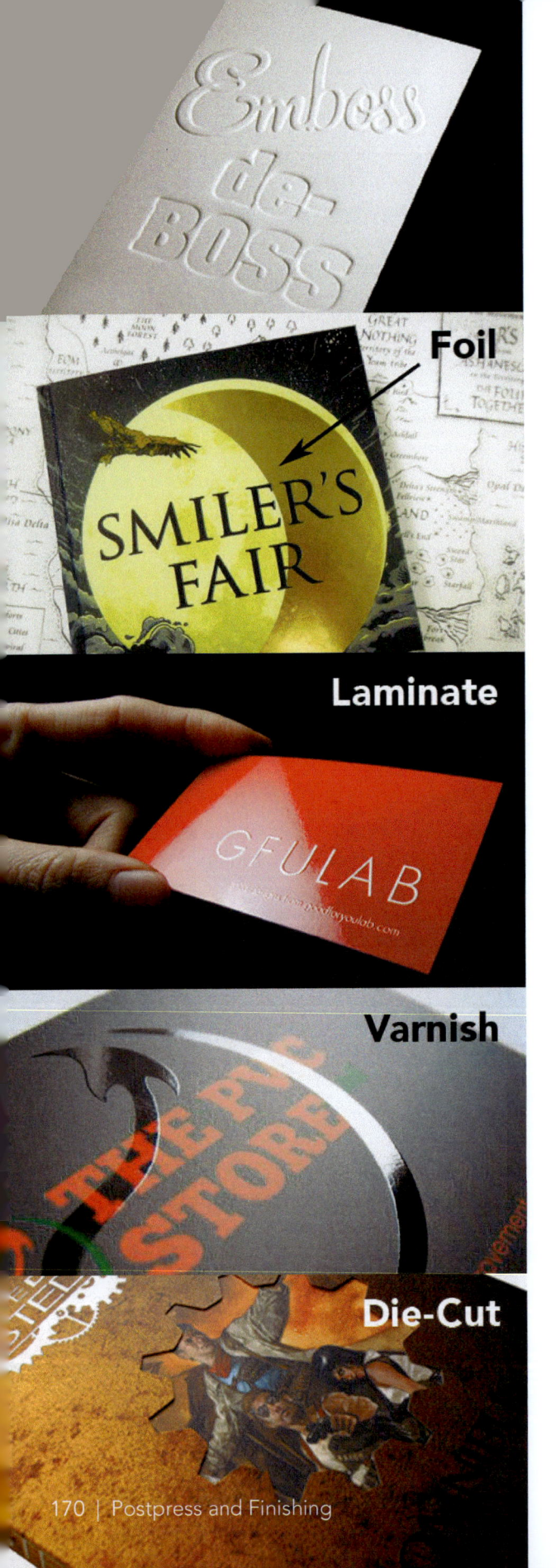

on one or more of data-specific considerations, including geographic location, income, or interests.

Demographic binding can be used in conjunction with Variable Data Printing or VDP (discussed in Chapter 8). Magazines with region-specific or interest-specific advertising can be delivered more easily to the right target audience. However, it is also used simply to reduce delivery costs, by sorting and segmenting shipments to National Distribution Centers for a postal rate discount.

Inline finishing equipment significantly reduces the labor cost for finishing operations while increasing the rate of production.

Special Effects

For high-impact pieces, special effects are the delight of graphic designers and, historically, a high-cost challenge for print professionals. They are applied at many points in the manufacturing process, both during printing, as with spot varnish effects, or in postpress. While some finishing effects require costly manual labor, an increasing number are now applicable with digital, inline technology.

Embossing creates a raised printed image whereas debossing creates a recessed image. Typically applied after printing, embossing is often used on paperback book covers, to accentuate the title or other visual element. Foil stamping can also provide a metallic look to the embossed or debossed image. "Blind" embossing or debossing creates a raised or recessed impression without a printed image.

Embossing and debossing effects are used not only to accentuate visual design elements, but also to give the paper surface an additional texture or dimensional effect. This too must be done after printing, which requires a flat surface for optimum reproduction quality. Metallic foils (foil stamping) and laminates are also applied after printing, for many of the same reasons.

Die-cutting is the process where a segment of the printed document is cut out of the page or sheet—using fixed metal blades or lasers—into a circle, triangle, or other shape. Book covers and greeting cards often feature innovative die-cutting and special foils or laminates to accentuate their appeal.

Die-cutting, embossing, debossing, and foil stamping are becoming increasingly popular due to the availability of automated processes for producing dies. Die-cutting has become increasingly used to embellish printed products due to automated electronic processes including CNC (Computer Numeric Controlled) machines. Such machines have allowed fine enhancements to printing for embossing, debossing, 3D die-cutting, and for foil stamping, and are increasingly being used for package printing, greeting cards, book cover enhancements, and for the general range of commercial printing. Programmed CNC machines eliminate the tedious hand-made dies traditionally used in the industry and are available from a number of manufacturers.

Packaging (Chapter 10) frequently takes advantage of special finishing effects, particularly with luxury products. Brand marketers and advertisers use these effects to accentuate the consumer's visceral connection with the brand.

Special effects are also used to enhance security and thwart counterfeiting of high-end products. Holographic images, lenticular effects, and other, striking images not only make a package stand out, they also make it hard to copy.

Special finishing effects are becoming highly automatable and integrated with other inline systems. The magazine cover pictured on page 165, for example has all the appearance of hand crafting. However, the combination of hot foil stamping and 3D spot coating were all digitally produced.

An inline finishing unit for a digital press. (Courtesy of Standard Finishing Systems.)

The Digital Bindery

With digital printing's growing popularity, many service providers find they now need to offer a variety of binding and finishing services. Commercial printers, publishers, direct marketers, and service bureaus can offer digital print services: shorter print runs; delivery on demand; variable, customized, one-to-one content; as well as high-quality color. However, these value-added services mean little if the binding and finishing applied to the printed products are not of sufficiently high quality.

Traditional binderies designed to work with offset presses may not have the range of finishing capabilities suited for digital press output. Therefore, printers that bring finishing processes into their plants may need to invest in a newer generation of bindery equipment.

Digital postpress has typically lagged behind developments in press technology. Increasingly, however, digital press manufacturers are increasingly working in partnership with finishing equipment OEMs to create new inline solutions for roll-fed and cut-sheet digital presses.

> *Cutting, folding, and binding are required for digital output, but on a different scale and with different workflow requirements.*

Digital postpress has unique challenges. Inkjet and electrophotographic (toner-based) presses typically work with shorter press runs involving quick turnaround and greater customization, such as the use of Variable Data Printing. Paper for digital print has different characteristics, and printed sheet sizes are typically smaller than those of conventional offset or gravure presses. Cutting, folding, binding, and finishing are still required for digital output, but on a different scale and with different workflow requirements.

As with all other aspects of print, postpress is increasingly subject to the needs of automation and supply chain management. To facilitate such control, between multiple vendors of equipment and management systems, a common, standardized approach is needed.

The Modern Job Ticket

Several print technology companies have proposed a unified approach to electronic job tickets—the digital equivalent to a printed job envelope. The latter hold the important documents and components of a job (such as samples of previous versions) and was printed with blank spaces for instructions and sign-offs needed at each step. The digital initiative became a recognized standard—the Job Definition Format or JDF—and is covered in greater detail in Chapter 11.

Although highly applicable to conventional printing, JDF has been somewhat less so with respect to certain aspects of digital printing, and especially to the inline complications of digital postpress operations.

Optimally, as part of a JDF workflow, job parameters are specified up front in the job ticket but can shift rapidly for variable-data printing. Also, control must extend to the equipment involved and to intelligence about the state of the product in the binding process itself—such as page order and number, insertions, covers—while handling errors that may occur. This can be done through direct feedback within inline solutions, or by using bar codes or related marks.

Other idiosyncrasies of digitally printed pages impact the effectiveness of automated, JDF-driven postpress workflows. These include special handling requirements for toner-based output, and the tendency of some devices to

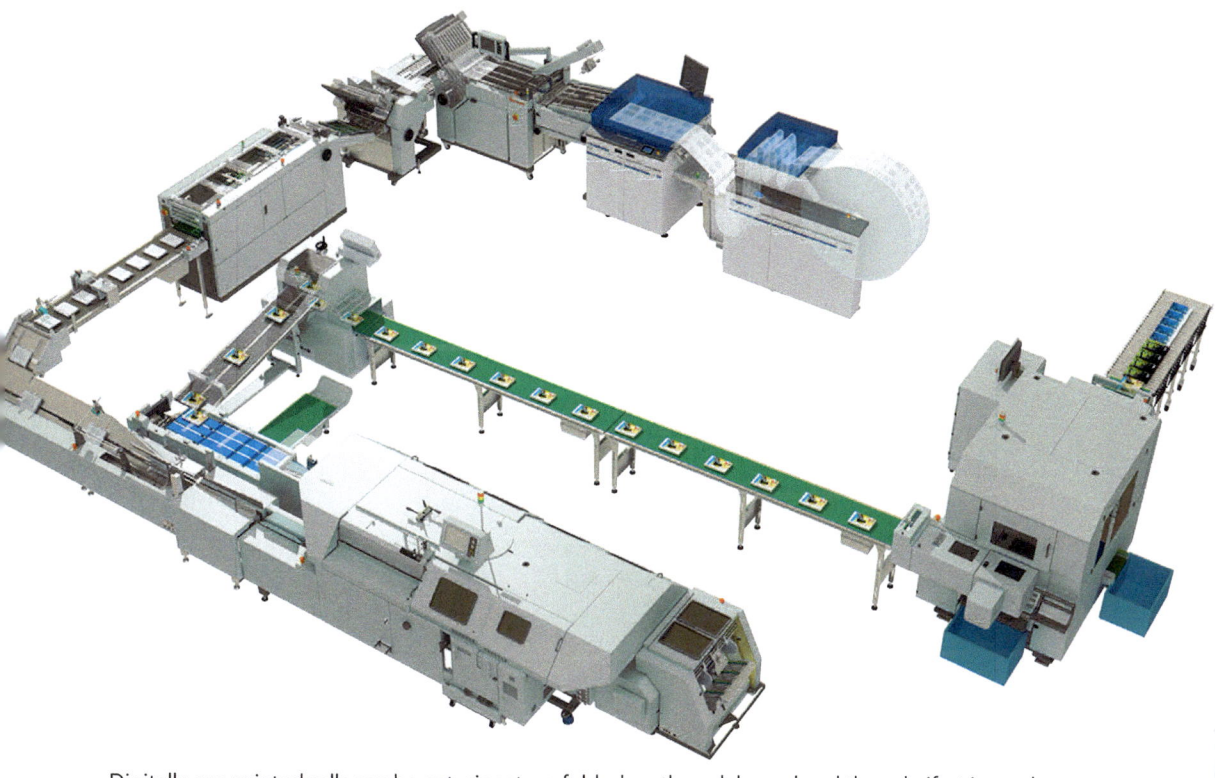

Digitally pre-printed rolls can be cut, signature-folded, gathered, bound and three-knife trimmed on a fully-automated binding line. (Courtesy of Horizon and Hunkeler.)

"drift"—as much as 1.5mm in any direction—when compared to offset. Similarly, digital printing presses tend to leave residues of silicon on the page. This residue can cause poor glue adherence when gluing the spine of a book block. Therefore, adequate milling and notching tools must be an intrinsic part of a digital perfect binding process. Finally, because digital pages can cling to each other because of static in a sheetfed bindery, an antistatic function in the bindery feeder will be critical to ensure the pages are fed one by one.

Facing the Future

As digital print suppliers approach their digital press and finishing vendors for help in deciding which bindery to use, uppermost in their minds should be the kind of press they have, the job they do, and the level of communication they require between press and bindery. Many digital press manufacturers and bindery vendors have varying degrees of "digital optimization."

Until now, digital press vendors have provided proprietary answers, using their own control languages. When mixing and matching, however, it is too expensive for finishing vendors to be writing new code for every digital press to which they may want to connect.

Recently, with the growing demand for integrated (inline) or semi-integrated (nearline) finishing, it became clear that a new open communications standard was needed to allow digital presses and finishing devices to talk to each other.

In response to these issues, OEMs of digital presses and postpress equipment developed the Universal Printer, Pre- and Post-Processing Interface or UP3i.

Compatible with JDF, it is an attempt to solve problems inherent with digital devices. These include:

- Shortening setup and prep times
- Enabling remote setup
- Maximizing the efficiency of any printing or finishing line
- Facilitating implementation with job tickets (especially JDF)
- Supporting both continuous form and cut-sheet printers
- Enhancing job recovery in the case of common mechanical occurrences, such as paper jams

Most sophisticated binderies work in tandem with other modules from the same or different vendors, including feeders (both roll-fed and sheetfed), rewinders (for roll-fed), creasers, trimmers, and collating towers. For perfect binders, they work with cooling towers for hot-glue solutions, as well as laminating devices for cover finishing. Even stackers and packagers may be part of the equipment mix.

Similar to offset binderies, digital bindery solutions are generally optimized to work with roll-fed or cut-sheet presses, as with black-and-white or color output, though bringing monochrome and color pages together is often a key part of the bindery process. Electronic binderies also span a range of types from loose bound (spiral, wire, plastic, and ring) to stitched (thread or staple saddle stitching-centered on the spine, or alternately, side or corner stitching), extending to perfect-bound and hardcover. Saddle-stitch and perfect-bound binderies are the primary focus of digital finishing vendors today, though "digital book factories" may also include solutions for hardbound books.

Over time, UP3i compliance will allow print service providers to add inline finishing capabilities to their growing digital footprint, without incurring the costs of proprietary interface connections between systems. In an environment with multiple system vendors and thin profit margins, this will improve the odds of business success.

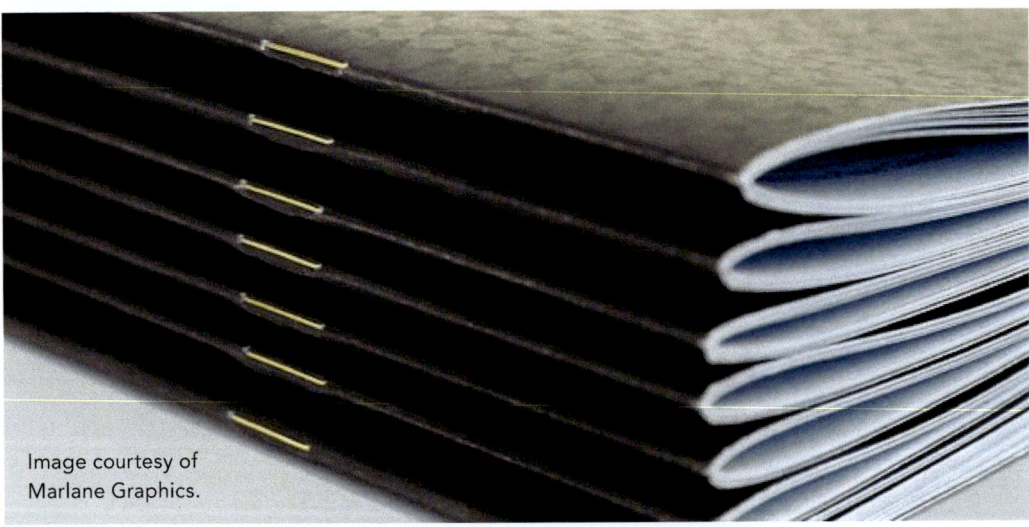

Image courtesy of Marlane Graphics.

10
The World of Packaging

The World of Packaging

Chapter Preview

The One-To-One Packaging Connection

Industry Growth

Types of Packaging

Digital Advantages

Market-Driven Trends

Interactive Media: www.igcvideo/Chapter10   Web Links: www.igcbook/Chapter10

Overview

Packaging has grown to become one of the largest segments of the graphic communication industry. Of the 36 related industry segments, 24 are classified as service providers. As noted in Chapter 5, these include a range of businesses—some changing rapidly but still viable, and others on the verge of extinction. However, one particular group of graphic communication service providers has not been negatively impacted by the Internet, and is outpacing the growth of all others. That segment is product packaging.

The growth and durability of packaging is easy to understand. There is simply no Internet replacement for it. Packaging consists of cartons, labels, flexible containers, and other printed, physical objects that contain and describe other physical objects. Consumer products and the packages that contain them have an intrinsic value in the world's market-driven economies. The Internet can describe them and facilitate their production, but is not a substitute.

The expanding societal and economic value of consumer products is nothing new. It is the reason why nearly every digital press OEM is focusing on packaging. The Smithers Pira report, "The Future of Digital Print for Packaging to 2022," describes the packaging trend, and its increasingly digital nature. As populations increase in size, and consumers in industrialized regions enjoy growing income, packaging demand rises. North America and Europe/Middle East/Africa (EMEA) are the largest markets for color digital label and packaging presses. Asia Pacific

Introduction to Graphic Communication | 177

and Japan, and the rest of world, are the smaller but faster-growing regions. In more developed regions, average household sizes are falling, reducing the size but increasing the complexity of average print runs. This plays to the strengths of digital packaging.

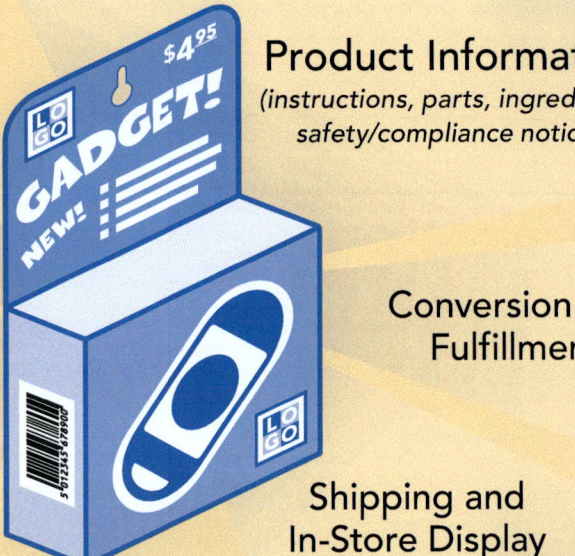

Packaging is Complicated

Brand Identity and Marketing

Logistics Support
(RFID, barcodes)

Product Information
(instructions, parts, ingredients, safety/compliance notices)

Conversion and Fulfillment

Shipping and In-Store Display

Packaging is a uniquely complex process for graphic communication professionals. A single product package must satisfy the needs of brand owners, marketing, compliance, production, fulfillment, logistics, and more. It must be an effective visual message, an accurate descriptor, an efficient container, and a practical inventory mechanism. Above all, it must satisfy the overriding concern of print packaging buyers: reduced time-to-market.

This chapter will outline the complexity of packaging, and the technology required to produce it efficiently. It will also explore the opportunities for commercial printers and others to expand into this profitable industry segment. For certain types of packaging, digital technology has opened up what was formerly a highly-specialized, niche market.

New Designations and Metaphors

The growth and viability of package printing was evident at DRUPA 2016, designated "The Digital Packaging DRUPA." Production inkjet and toner equipment manufacturers showed numerous new technologies for digital package printing, particularly for folding cartons, corrugated boards, and labels.

The "direct-to" designation has been applied to the modern printing industry for decades. "Direct-to-plate" and "direct-to-press" are two examples from Chapter 9. Two new ones are "direct-to-board" corrugated printing—for printing on already pre-formed corrugated boxes—and "direct-to-shape." The latter refers to customized or variable data printing (VDP) directly on a pre-formed package of any type. Increasingly, this is targeted to select demographic groups.

> A typical supermarket shopper spends only 2.6 seconds on the decision to pick up a package.

The trend towards personalization and segmentation, already a major consideration in the general printing industry, is especially relevant in packaging. A recent study found that a typical supermarket shopper spends an average of only 2.6 seconds making the decision to pick up a package from a shelf. With 60 percent of purchases described as "impulse purchases" and 80 percent of all decisions being made in the store, the impact of customized or personalized packaging cannot be underestimated.

With high-speed, automated production of increasingly short runs, augmented by VDP and other digital technologies, the new metaphor for packaging is a one-to-one, personal connection between the product manufacturer and the individual consumer.

Packaging Industry Growth and Worth

The packaging industry is growing rapidly, and does not show signs of slowing down any time soon. According to the Smithers Pira report, the estimated worth of digitally printed packaging was $13.2 billion in 2017, and forecast to reach $23.2 billion by 2022.

The forecasting group MarketsandMarkets estimates that the global packaging print market will grow from $120 billion in 2016 to almost $193 billion by 2026.

Printing Industries of America predicts that the U.S. packaging industry will expereience growth of over 12 percent from 2016 to 2024.

Labels are predicted to be the largest segment during the forecast period, due to applications such as displaying sequential barcodes and numbering, variable text, titles, or graphics. However, folding carton, flexible packaging, and corrugated board applications are also growing rapidly through digital printing.

According to Smithers Pira, analog packaging is also growing by about 28 percent per year. This growth is expected to continue. In contrast, however, all *digitally* produced packaging is expected to increase by over 300 percent in the next few years. This expected growth explains why many OEMs are developing digital equipment for the main packaging industry segments.

Digital Inkjet and Electrophotography

The three traditional processes for package printing—flexography, gravure, and lithography—will continue to be vital and necessary, particularly for static layouts and long-run printing. However, their market share will gradually diminish as digital printing processes advance in speed and flexibility, and as print runs become smaller and more customized.

> Some types of packaging—such as labels and folding cartons—are a better fit for commercial printers than highly specialized ones.

Digital printing is already well established in label production and other segments, with electrophotographic (EP) and inkjet enjoying growing shares of the market. These processes are expected to continue growing. In 2010, these accounted for $166 million in sales of color digital presses alone, and have been growing by over 10 percent a year since then.

EP and inkjet are improving in quality and speed. Larger digital systems and presses are becoming available, challenging the dominance of conventional presses. Converters (companies that turn printed sheets or rolls into finished packages) now frequently use color digital presses in-line with their highly automated production systems. In many cases, this includes not only assembling and finishing the package but also inserting the product and assembling/preparing multiple products for shipment.

Nearly all digital press OEMs are developing EP or inkjet systems—and related technologies—for packaging workflows. Some of these include special inkjet heads, web delivery systems, in-line wrapping and finishing, and 3D visualization software for creating and reviewing package design.

Many of these digital print technologies are well within the reach of commercial printers, who can use them to expand their business from general-purpose printing to packaging applications. The growth of packaging in general makes this an appealing financial prospect.

However, some types of packaging—such as labels and folding cartons—are a better fit for commercial printers than highly specialized ones, such as flexible packaging and rigid plastic or metal. In addition to their dissimilarity to ink-on-paper printing, these specialized segments are very often incorporated by converters and co-packers—with whom commercial printers could not easily compete.

Packaging Industry Segments

Folding Cartons

As the name indicates, this involves printing on thick, paper-based substrates, which are subsequently folded and glued to form the package. This poses challenges to the designer, who must anticipate the finished, three-dimensional product.

Typically, the graphic designer works in cooperation with a computer-aided design (CAD) specialist, who supplies the non-printing fold and die lines. Designers must take care not to place critical elements too near these elements, and avoid areas such as glue tabs.

Although folding cartons are typically printed on one side of the sheet, special imposition must often be used—nesting items on the printed sheet to avoid waste.

The folding carton market is particularly well suited for digital printing—and for inclusion by commercial printing companies. Historically, nearly all folding cartons have been produced using conventional, offset printing. Therefore, as press runs become shorter, more customized, and even personalized, it is ideal for digital processes, especially inkjet.

Most folding carton applications involve single-sided printing on paperboard stock, which most presses can handle easily. When the printed sheets are delivered to co-packers, there is little or no concern about contact with food.

Hence, the B2 (19.7 x 27.8 inch) sheet size typically used for commercial printing is also well suited for folding cartons.

Digital presses used for conventional and folding carton printing include those made by Canon/Océ, HP, Landa, Ricoh, Screen, and others.

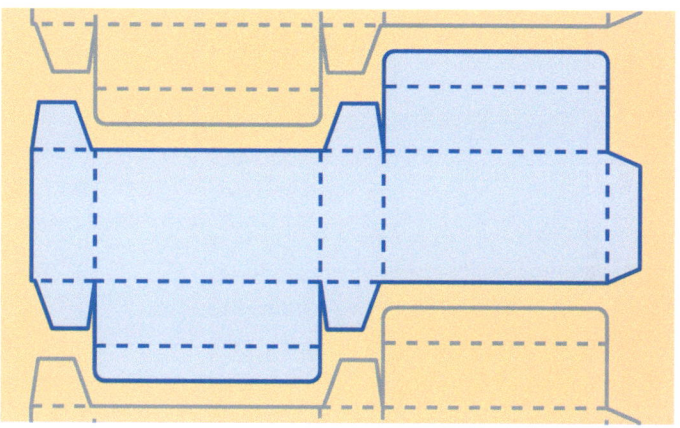

Labels

Product labels can be simple and utilitarian or adorned with elaborate colors, coatings, and special effects. As with other packaging, labels frequently carry a market-driven visual message as well as essential product information.

Labels are frequently printed in multiples, on a narrow-width web press. Although this was traditionally done on flexographic presses, it is increasingly moving to digital inkjet devices. Press manufacturers for this segment include HP and Xeikon as well as Allen Datagraph, Primera, Quick Label Systems, and others.

Label printing is often done on hybrid or combination printing systems, where printing and finishing is all done in a single pass. Some presses print at speeds up to 262 feet (80 meters) per minute.

Due to stringent labeling requirements, the use of late-stage customization to print the final details of a package or label are being used more frequently—a trend led by the pharmaceutical industry.

Nearly all labels are one-sided, with some exceptions for labels on glass or other transparent containers. What makes them distinctive—and problematic for printers—is the need for an adhesive surface on the back. Label stock is typically a layered substrate, with a self-adhesive coating and a protective layer, which is removed when the labels are applied to the container.

Labels must also be die cut during the finishing process. This imposes requirements on the designer—to keep critical elements sufficiently far from the edges. It also requires an efficient process for removing the excess material, often at high speeds.

Flexible Packaging

Printed on paper, plastic, film, and metal foil, flexible packaging is used for a wide variety of consumer products. It includes bags, pouches, labels, liners, wraps, and other containers—typically for food products, or anything that requires moisture, temperature, or portion control.

The flexible packaging industry is huge, profitable, and growing, representing over $50 billion in revenue internationally.

Flexible packaging is traditionally produced using flexographic presses, which could easily handle non-paper substrates at high speeds. Such presses are often incorporated in full production lines, where the actual product was made, portioned, packaged, and prepared for delivery. Digital inkjet devices are starting to make inroads into this sector, but flexible packaging presents more issues for digital technology than folding cartons or labels. However, the industry is rapidly dealing with several issues.

To begin with, flexible packaging uses thin, unsupported film media that can be problematic for digital presses. Reverse printing (making the substrate color form the logo or other image) is often problematic. So too is surface printing—a method used on uneven surfaces. Additionally, since most of these packages are used for food products, toxicity concerns and FDA regulations must be met.

Despite this, brand interest in digital printing is strong, due in part to the need for cost-effective short runs and customization. Digital press OEMs entering the flexible packaging sector include Fujifilm, HP, Kodak, Landa, and others.

Flexible packaging poses its own challenges to graphic communicators. Designers must consider the various tabs, seals, distortion created by substrate shrinkage, and other aspects of the finished product. Even the order in which the package is folded and sealed can affect the final appearance. In many cases, 3D design software such as Esko's ArtiosCAD or Creative Edge's iC3D are used to better visualize the piece.

Finally, color management can be a significant issue for brands using flexible packaging. For both flexographic and digital printing, the color behavior of inks on plastic, film, and foil is often quite different from that of paper.

Corrugated

Corrugated boxes differ from folding cartons in several important ways. The laminated substrate is often made from recycled fiber. Designed for strength, it is utilitarian in nature, and often unsuitable for high-resolution graphics.

Corrugated boxes are typically used to ship products, individually or in quantity. They can also be used in point-of-sale displays or for generic packaging in warehouse retail stores. As such, they usually have much simpler folding requirements, and do not often require color printing.

Corrugated board printing represents an area of promise for high-end presses, because brands increasingly need short runs for their corrugated packaging OEMs competing in this area. Some include Barbaran, Bobst, CorrStream, Durst, HP Scitex, and Inca.

One of the expected growth areas for corrugated board printing is in direct-to-board inkjet printing. One example of such a system is Xanté's wide format printer, driven by a Memjet engine, that prints at a speed of 10- to 12-inches per second. A promising growth area for corrugated is inkjet printing on pre-formed corrugated containers.

Rigid Plastics

As a lightweight replacement for glass or metal, rigid plastic is a fast-growing packaging industry segment. Smithers Pira estimates growth from $166 billion in 2017 to over $200 billion in 2022. Primarily used for food and beverage products, it is also widely used for healthcare, personal care, and household products.

Traditionally done via screen-printing, today rigid plastic is commonly printed using inkjet—from simple, but sometimes variable product and date codes to brand-critical color graphics.

As with flexible packaging, rigid plastic is commonly thought to create a significant waste problem. The issue is a complex one, however. Industry advocates maintain that recycling initiatives offset the environmental impact, and that the use of rigid plastic lowers the energy cost of shipping by decreasing overall weight.

The great growth potential for rigid plastic packaging is "direct-to-shape" printing using variable data printing on pre-formed containers such as cups, cans, or other cylindrical containers where each piece can be personalized. Existing systems include Machines Dubuit, Sylvan Print Technology, and Tonejet.

Metal Packaging

In the past, "metal decorating" consisted of printing on sheets of metal *before* they were formed into cans for beverages and other products. This was formerly a large segment of the packaging industry, but is now no longer technically or economically viable.

Today, metal packaging is largely done in-line, using digital inkjet and other print processes to directly image the can or other container during the manufacturing and filling process.

Packaging's Digital Advantages

The packaging industry has relatively recently adopted digital technologies with advantages that have been enjoyed in general commercial printing for some time. To the general digital printing list, packaging adds one more: the "smart packaging" phenomenon. It is a development that promises to revolutionize not only how packages are produced, but also how they are used by retailers and consumers.

Short Run

The demand for short-run, digitally printed packages is rising. Digital presses have made it economically feasible to produce printed packages of runs of 5,000 or fewer. This allows companies to reduce inventory and storage costs, and produce printed packages in quantities more in line with the market lifecycle of their products. That time interval is often very short. Digital printing also facilitates quick and economical changes in product ingredient and nutrition markings for food products—allowing shorter time-to-market for new product variations.

Variable Data Packaging

The variable data capability of digital presses also allows for versioning and personalization that inspires rapid consumer responses. Variable data printing is typically identified with short-run printing. However, there are increasing examples of variable data packaging not being restricted to short runs, but used for long runs as well. A classic example of this is the highly publicized Coca Cola case study.

Entitled "Share a Coke," Coca Cola took advantage of the variable data capability of digital presses, and used versioning and personalization for a campaign in 35 European countries involving the printing of over 750 million packages. As part of its campaign, Coca Cola produced over a billion labels; putting the popular notion to rest that digital printing is only suitable for short runs.

The campaign increased social media engagement and sales by versioning and personalizing the labels to appeal to different demographics in different nations.

If a billion labels can be segmented and personalized for multiple nations, shorter print runs can be versioned and personalized for the demographics of different regions within a nation, and down to neighborhoods and special interests of specific residents. The ability to produce short print runs economically will grow in interest as packaging buyers continue to search for innovative digital methods and ways to engage with present and prospective customers.

Print On-Demand

Printing on-demand means less waste, ensuring that new designs, product details, or changes in ingredients do not result in redundant inventory. Technological developments in inkjet and electrophotography have simplified and made print on-demand realistic for packaging. It has made personalization of packages by region, special interests, demographics, and so on, practical for packaging.

The concept and promise of print on-demand package printing has inspired OEMs to focus their press technologies and marketing on the packaging industry. As previously noted, nearly every manufacturer of digital printing presses is focusing on this market. Interestingly, it is also a market with resources. Unlike other printing industry segments, packaging printing is growing, particularly for consumer-packaged goods (CPGs) and fast-moving consumer goods (FMCGs) produced by large companies that understand brand user and consumer preferences. They have the resources to invest in new technologies to help gain market share. The digital press manufacturers understand this.

Personalization

Personalization, or the ability to personalize data within a printed document, via variable data printing, has been an attribute of digital printing for the commercial printing industry sectors for a couple of decades. However, it is new to packaging, and promises to be a major application for label, folding carton, and flexible package printing, and will grow as a major marketing force for selling packaged products.

It is projected that personalized packages could potentially increase product sales from 100 to 1000 percent. Personalization could be by region, for seasonal events, group demographics, past purchasing experiences, and even down to the individual person through new technologies being developed for "Smart Packaging."

Smart Packaging

Another area where packaging has taken on a new role is in the area of "Smart Packaging." This is where images on a substrate can potentially communicate with the consumer and the consumer with the printed image.

The potential of digital printing is broadening, and packaging companies are only just beginning to realize the benefits offered of personalization, customer engagement, and integration with online campaigns and marketing at the point of sales. Smart Packaging, also referred to as "active packaging" or "intelligent packaging" refer to packaging systems used for foods, pharmaceuticals, and several other types of products. They help extend shelf life, monitor freshness, display information on quality, improve safety for the consumer, and improve shopping convenience.

Here is where through Near-field Communication (NFC), Quick Response Codes (QR), Universal Product Codes (UPC), and Radio Frequency Identification (RFID), the relevance of a package's information such as nutrition, heath, and wellness information can change instantaneously, depending on the consumer who looks at or purchases the product. Advantages of the product to the consumer, as well as warning and cautions, would appear that are relevant to the specific consumer.

As one example, CVS Pharmacy developed a promotion called "In-store Alerts." Through a mobile device with a CVS app, installed, the consumer can receive notification alerts right in the store, announcing special deals on products of possible interest to consumers based on previous purchasing histories.

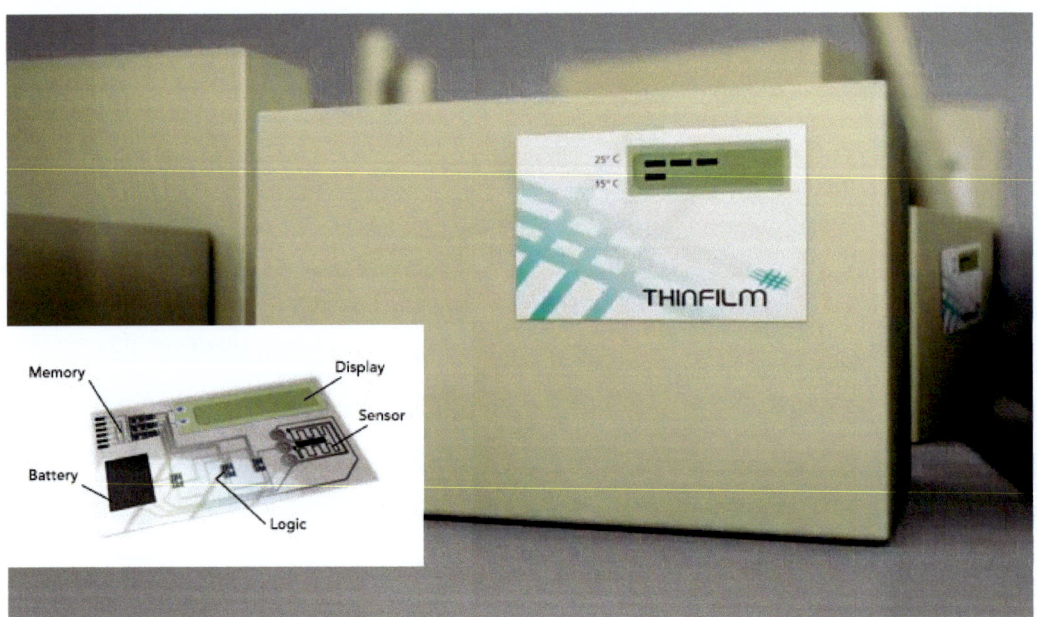

Smart labels can provide a wide range of vital information, including environment conditions during the product shipping process. (Image courtest of Thinfilm.)

CASE STUDY

An example of one of the latest packaging "start-ups," using the value of digital packaging, is ePac LLC in Madison, Wisconsin.

In its first year of operation, ePac, focused exclusively on flexible packaging, has committed to maintain entirely digital applications from front-end through printing in providing a focus on high speed, short-runs, rapid replacement of package turnover, and minimum inventories for its clients.

Carl Joachim, Chief Technology Officer for ePac, shared that in April of 2016 ePac began commercial operations with a technology platform based on the HP Indigo 20000 press. The company's applications, including prepress technology, printing, laminating, and pouch making, uses an end-to-end digital workflow for automated production processes. ePac has already installed a second HP 20000 in its Madison location, and has committed to acquire additional presses to fuel planned growth in North America. Further, Joachim added, work is underway to develop a fully automated platform to mechanize the steps from order entry through file preparation and printing.

Jack Knott, ePac's CEO, said, "By using digital applications, we built ePac so any company, large or small, can turn around products quickly on the retail shelf, decrease inventory and inventory obsolescence, and grow their top lines. ePac was built to help customers access modern technology for marketing and sales, to reduce the environmental footprint of the process, and to realize the value of digital processes for packaging purposes."

From its initial planning, the intent was that ePac would be an anomaly and a seminal company in demonstrating that a flexible package printer can rely on a total digital workflow with no conventional flexographic or gravure presses. ePac's clients include several major consumer packaged goods companies (CPGs). The demand for digital packaging has grown so rapidly that ePac has opened two additional facilities, one in Boulder, Colorado and one in Los Angeles, California.

Press Cost Considerations

Traditional press manufacturers (OEMs of analog gravure, flexographic, and offset lithographic presses) focusing on the packaging industry should take notice of an interesting development. Although the advantage of digital presses has created a viable short-run market for packaging, traditional press manufacturers continued to see the need to produce analog presses to serve long runs. Such presses are very expensive, bringing in millions of dollars in revenue to the press manufacturers. Indeed, these traditional presses, many comprised of numerous printing stations beyond four units, can cost multiple millions of dollars. However, this dynamic is changing with digital presses now being developed to serve the long-run packaging market.

Million-dollar, high-end presses are now redefining the role of digital printing in packaging. The number of such packaging presses is growing internationally. All of them are designed to print larger image areas on larger paper sizes for folding cartons and flexible packaging.

> **The introduction of costly digital packaging presses will bring about a fundamental market shift.**

Further, there had been the development of high-end digital printing presses that print directly to corrugated board in full-color, and on cans, bottles, tubes, and other packaging shapes. These are all color digital presses, but with prices of between $1- and $4-million. Hence, the market for long-run packaging now has "digital options" not previously available, and represent competition and a threat to OEMs of traditional analog presses for packaging.

The output speed and productivity of such presses have grown to compete with traditional presses as well. For example, the 1.8-meter-wide EFI Nozomi C18000 LED inkjet press can output up to 9,000 boards per hour. The Konica Minolta KM-C inkjet press can output up to 2,200 cardboard and micro-flute boards per hour.

Some press developments have come from industries outside of the printing industry. This includes the beverage industry and similar product companies seeking greater packaging automation.

The introduction of costly digital packaging presses will bring about a fundamental market shift. Placements will be fewer, compared to digital label presses, but growth is expected. To justify their cost, output will far exceed the package printing volume that a short-run digital label press can achieve. Target products for these presses are expected to be bottles and corrugated boxes. Such presses will also appeal to large companies with revenue of $10 million or more that are new to creating production-level digital printing for packaging. This includes can manufacturers and folding carton converters. It is expected that higher-performance digital presses will enter the market at an accelerated rate.

Some manufacturers of such high-end presses include Barbaran, Bobst, CorrStream, FujiFilm, HP Scitex, HP Indigo, KHS, Konica Minolta, Landa, Mark Andy, Océ, Omet, Screen, and Till.

Paper Considerations

Related to press costs are paper considerations. Digital presses supporting the B2 size format (19.7 x 27.8 inch or 500 x 707 mm) are starting to make their presence felt in corrugated, folding carton, and flexible packaging, where analog production still dominates. Although placement of larger format digital presses is currently modest, interest and installations continue to grow. For example, there has been a recent increase in the installation of the 30-inch format size by folding carton printers along with the installation of "half size" analog high-productivity presses.

Finishing and Lamination

Finishing processes are very important in producing the complete package. Therefore, with advances in digital press development for packaging must come advances in finishing processes. Finishing is vital for many reasons. It influences package design, placement of products in packages, storability of packages in stores and in the home, food product safety concerns, and how packaging influences consumer purchases.

Lamination, coating, die cutting, spot varnishes, folding, and scoring are vital capabilities required in effective package production. Increasingly, these are being combined and automated, either by individual OEMs or through partnerships.

For example, the TRESU iCoat 30000 now offers protective and spot varnish in one pass as well as new enhancement capabilities with metallic and other high-viscosity, flexographic inks. The HP Indigo 30000 is also compatible with partner converting solutions for inspection, creasing, folding, and gluing.

Another example is the HP Indigo Pack Ready laminator. This creates a film that does not require curing, so it connects directly to an in-line Form, Fill, and Seal (FFS) machine to produce pouches with no lag time. The lamination process on the Pack Ready laminator combines different films, each providing a different functionality (print quality for shelf appeal, protection of food for longer shelf life, sealant film to melt together via a heating process), just as converters do. The quality of the bond strength between substrates being joined is key, as lamination failure could lead to compromised food. (Until recently, traditional, solvent-based lamination for flexible packaging, required a lengthy curing process that precluded an automated, in-line process in food packaging.)

In other words, advances in finishing and lamination technology, paralleling those of digital printing, will continue to transform the packaging process worldwide, by making it faster and more automatic.

Conclusions

Packaging is a growing industry segment, not negatively impacted by digital media. The purpose of this chapter was to show how general commercial printers and other graphic communication companies are entering it. The chapter also provided an overview of digital package printing, its applications, and advantages.

> *Market-driven interests and values are not likely to change, nor will the growth of consumer product packaging.*

The growth and success of the consumer packaging industry is related to the market-driven interests and values of Europe and North America, and the expansion of these values in Asia and elsewhere. These interests and values are not likely to change, nor will the ongoing growth of the consumer product packaging industry in the decades ahead.

11

Best Practices and Industry Standards

Best Practices and Industry Standards

Chapter Preview

Job Specifications

Elements of a Print Transaction

Standards and Specifications

Overview

The term "business practices" as presented here refers to the common practices of the printing industry. Business practices are recommended guidelines and are flexible enough to accommodate various client and service provider relationships. These business practices are generally recognized in credible industry publications and endorsed by leading industry professional associations.

It is recognized in today's competitive environment that the most successful printers and print brokers offer their clients a full range of products, exceptional support services, first-rate customer service, timely delivery, and quality printed products.

The Business Process

The relationship between the printer or print broker and the client cannot be at odds. Teamwork is key to keeping customers happy as cycle times are reduced and technology constantly shifts. Achieving this requires a commitment from each team member to work toward a unified group of objectives and dedication to supporting each other's actions.

The print buyer or content creator is not always familiar with all facets of the printing process and printing production and, therefore, places trust in the printer or broker to provide fair and honest assessments of what it will take and cost to produce and deliver a job that meets the requirements.

The Primary Duties of Printers, Print Brokers, and CSRs

Printers and print brokers typically rely on a customer service representative (CSR) to coordinate details of work-in-process. The CSR functions as a liaison between the printing company and customer. The CSR's job involves coordinating activities between estimators, sales representatives, production management, and other internal printing company staff. It also involves facilitating timely delivery of each customer's job, establishing, and maintaining an agreed-upon level of quality. The CSR provides backup support for sales, estimating, and production management, and serves as the "face" of the company. In this role, the printer or print broker is expected to provide the printed job at the highest quality, as quickly as possible, and at the lowest price—with minimal to no hassle to the print buyer.

Job Specifications

In printing, job specifications—often referred to as "specs"— are the key parameters that define an order. The printer or print broker is expected to assist the client in developing proper and accurate job specifications. Accuracy of specifications is very important, as this is what price quotations are based on. The service provider can re-quote a job at the time of submission if any of the specs change or if copy, film, tapes, disks, or other input materials do not conform to the information on which the original quotation was based.

It is also important that the service provider (such as the printer or print broker) works with the client to prepare a professional Job Specification Form (JSF). This is a form that the printer or sales rep completes in discussion with the client. The purpose of this form is to provide essential information that describes the job accurately and fully. From this information, the job can be planned and economically evaluated.

Some customers provide their own written specifications. However, these documents are often not thorough or detailed enough because the print buyer is not always aware of the types of details that should be included. That is why, in such cases, it is ethical on the part of the printing company or print broker to help the client develop specific and accurate job specifications.

The printing company or print broker should make sure that the client is not rushed in developing job specifications. Rushed or incomplete specifications typically result in costly changes and disappointments later on.

The Elements of a Print Transaction

There are various elements in a print job from start to finish. They include providing clarity and agreement between the print buyer and the service provider—the printer. It is prudent to include each requirement in the final contract. Doing so will eliminate misunderstanding and, hopefully, disputes during the production

Typical Information on a Job Specification Form

- The customer's name, address, and telephone number
- The sales representative's name
- A general description or the job
- Quantity levels
- Flat sheet and folded dimensions (with possible alternate sizes)
- Bleeds and margin information
- The number of colors
- Complete details regarding all artwork (who will supply it and whether it will be ready for prepress such as scanning, or digital input)
- Complete paper specifications
- Binding, finishing, and other postpress requirements
- The type of packaging to be used
- The type of delivery method and time of delivery

of the print job and after its completion and delivery. The elements include:

An Estimate is generated and used by the printing company as an internal dollar measurement of the cost required to produce a specific product. A cost estimate is an approximation and is generally not shared with the customer. It does not necessarily represent the final cost.

A Proposal is a tentative offer to produce a printed product for a specific dollar amount. It is not the final cost and is subject to negotiation and change. The proposal is a formal, typewritten document that specifies price for a printing order. A complete proposal describes the work to be done, the expected levels of quality and quantity, and provides tentative selling prices.

A Quotation is an offer from a printer or print broker to a customer to produce a printed product for a specified dollar amount. Unlike a proposal, the quotation is final and binding to both parties. Once the price quotation has been agreed upon, it has the impact of a legal agreement.

Because the quotation is a legal contract, prudent printers and print brokers include with it a copy of the latest Business Practices for the Printing Industry. These practices were researched and drafted by the Printing Industries of America (PIA) and the National Association for Printing Leadership (NAPL) through surveys and feedback from hundreds of printers.

Typically, a quotation not accepted within thirty days may be changed.

The Order – Acceptance of orders is subject to credit approval and contingencies such as fire, water, strikes, theft, vandalism, "acts of God," and other causes beyond the service provider's control. Credit approval documents are ordinarily separate from customer agreements to acquire services. Canceled orders require compensations for incurred costs and related obligations.

Alterations/Corrections and Reprints – Customer alterations include all work performed in addition to the original specifications. All such work is subject to additional charges at the service provider's current rates.

> **The most frequent "mark-up over cost" used by commercial printers is 35%**

When a customer requests a straight reprint order, there is no rework necessary. Therefore, the manufacturing costs of the job are lower than in the initial run because prepress manufacturing has been completed. When a customer requests a minimum change reprint, the estimator must determine the quantity of rework necessary, then add its costs to the existing press and finishing costs. In either situation, because all or most of the prepress work has been completed for the reprint and is paid for by the customer on the first printing, manufacturing costs are reduced.

With this in mind, printing companies or print brokers provide a reprint discount on the billed price of the first printing. Discount rates and discounting vary. However, a range of 5% to 30% off of the initial price of the first job is common.

Over-Runs or Under-Runs usually do not exceed 10% of the quantity ordered. The service provider will bill for actual quantity delivered within this tolerance. If the customer requires a guaranteed quantity, the percentage of tolerance must be stated at the time of quotation.

Production Planning or job planning is the evaluation of manufacturing methods to ensure that the customer's order is processed to meet the job requirements. For planning to be thorough, it should be done carefully, and it should be based on the set of accurate and complete job specifications provided by the customer.

Production schedules will be established and followed by the customer and the service provider. In the event that the customer does not adhere to a production schedule, delivery dates will be subject to renegotiation. There will be no liability or penalty for delays due to state of war, riot, civil disorder, fire, strikes, accidents, action of government or civil authority, "acts of God," or other causes beyond the control of the provider. In such cases, schedules will be extended by an amount of time equal to the delay incurred.

Customer Credit should be checked prior to extending a proposal or quotation. However, this is not always possible in an environment of quick turnaround. In such cases, a statement such as "prices quoted are conditional upon company approval of customer's credit" on the

proposal or quotation clarifies the company's credit checking process. Credit check documents and quotations or proposals are separate documents; they should not be one and the same document.

There are three ways to check a potential customer's credit. One is to have the customer complete a credit application that requests credit references. A second is to contact a credit association. A third way is to use the credit referral service offered through most PIA local affiliates.

Markup – The most frequent "markup over cost" used by commercial printers is 35%. Thus, a job with a total cost of $100 would be priced at $135. However, print broker markups can range from 3% to 35%.

Invoicing – The customer should be invoiced as soon as possible after delivery of the printed goods. One working day or sooner is recommended; on the day the job is shipped is ideal. Thus, the customer receives the goods and immediately thereafter the invoice (with payment due within 30 days). Discounting for early payment is sometimes offered—i.e., 2% net 10 day.

Experimental or Preliminary Work performed at a customer's request will be charged to the customer at the service provider's current rates. This work cannot be used without the service provider's written consent.

Customer's Property – The service provider will only maintain fire and extended insurance coverage on property belonging to the customer while the property is in the service provider's possession, and liability for this property will not exceed the amount recoverable from the insurance. Additional insurance coverage may be obtained if it is

requested in writing, and if the premium is paid to the provider.

Creative Work – Sketches, copy, layouts, comprehensives, and all other creative work developed or furnished by the customer is the customer's exclusive property. If the service provider develops such work, it remains the property of the service provider, and the creator must give written approval for any and all use of this work and for any derivation of ideas from it. In sum, the use of artwork, type, plates, negatives, positives, tapes, disks, and all other items remains the exclusive property of the provider of such items.

Electronic Manuscript or Image – It is the customer's responsibility to maintain a copy of the original file. The service provider is not responsible for accidental damage to media supplied by the customer or for the accuracy of furnished input or final output. Until the provider can evaluate digital input, no claims or promises are made about the provider's ability to work with jobs submitted in digital format, and no liability is assumed for problems that may arise. Any additional translating, editing, or programming needed to use customer-supplied files will be charged at prevailing rates.

Customer-Furnished Materials and Purchases – Material furnished by customers or their suppliers should be verified by delivery tickets. The provider bears no responsibility for discrepancies between delivery tickets and actual counts. It is up to the printer or print broker to make sure that all materials and supplies received are accurately documented. Customer supplied paper must be delivered according to specifications furnished by the service provider. These specifications will include correct weight, thickness, finish, size, and other technical requirements.

Artwork, film, color separations, special dyes, tapes, disks, or other materials furnished by the customer must be usable by the provider without alteration or repair. Items not meeting this requirement will be repaired by the customer, or by the provider for an additional charge. Unless otherwise agreed in writing, all outside purchases as requested or authorized by the customer are chargeable.

Prepress Proofs – The service provider will submit prepress proofs along with original copy for the customer's review and approval. Corrections will be returned to the service provider on a "master set" marked "Okay," "Okay with corrections," or "Revised proof required," and signed by the customer. Until the master set is received, no additional work will be performed. Approvals in writing will be provided for virtual or "soft proofs" viewed on a monitor as opposed to hardcopy.

The service provider will not be responsible for undetected production errors if proofs are not required by the customer, the work is printed per the customer's okay, or if requests for changes are communicated verbally.

Press Proofs will not be furnished unless they were required in the quotation. A press sheet can be submitted for the customer's approval as long as the customer is present at the press during makeready. Any press time lost, or alterations/corrections made, because of the customer's delay or change of mind will be charged at the service provider's current rates.

Color Proofing – Because of differences in equipment, paper, inks, and other conditions between color proofing and production press department operations, a reasonable variation in color between color proofs and the completed job is to be expected. The expected variation should be explained to the customer and the customer should acknowledge an understanding of this. Variations of this kind are considered acceptable.

Delivery – Unless otherwise specified, the price quoted is for a single shipment, without storage. Proposals are based on continuous and uninterrupted delivery of the complete order.

It is the responsibility of the customer to maintain an original copy of an electronic manuscript, page layout, or image file.

If the specifications state otherwise, the provider will charge accordingly at current rates. Charges for delivery of materials and supplies from the customer to the service provider, or from the customer's supplier to the provider, are not included in quotations unless specified. However, the customer should be made aware of the approximate cost once quantities and other job specifications have been determined. Also, once quantities and other job specifications have been determined, the service provider should make the customer aware of expected delivery charges after the job is completed.

Title for finished work passes to the customer upon delivery to the carrier at shipping point; or upon mailing of invoices for the finished work or its segments, whichever occurs first.

Storage – The service provider will retain intermediate materials until the customer has accepted the related end product. If requested by the customer, intermediate materials will be stored for an additional period at additional charge. The provider is not liable for any loss or damage to stored material beyond what is recoverable by the provider's insurance coverage.

Taxes – All amounts due for taxes and assessments will be added to the customer's invoice. Tax exemptions won't be granted unless the customer's "Exemption Certificate" or other proof of exemption accompanies the purchase order.

Telecommunications – Unless otherwise agreed, the customer pays for all transmission charges, and the service provider is not responsible for any errors, omissions, or extra costs resulting from faults in the transmission.

Terms/Claims/Liens – Payment is net cash thirty calendar days from date of invoice. Claims for defects, damages, or shortages must be made by the customer in writing no later than ten calendar days after delivery. By accepting the job, the customer acknowledges complete satisfaction.

Liability – In a Disclaimer of Express Warrantee, the service provider warrants that the work is as described in the purchase order.

The customer understands that all sketches, copy, dummies, and preparatory work shown to the customer are intended only to illustrate the general type and quality of the work and that they may not be intended to represent the actual work performed.

In a Disclaimer of Implied Warrantee, the service provider warrants only that the work will conform to the description contained in the purchase order.

The provider's maximum liability will not exceed the return of the amount invoiced for the work in dispute. Under no circumstances will the service provider be liable for specific, individual, or consequential damages.

Indemnification – The customer agrees to protect the service provider from economic loss and any other harmful consequences that could arise in connection with the work. This means that the customer will hold the provider harmless on any and all grounds. This will apply regardless of responsibility for negligence.

Copyrights – The customer also warrants that the subject matter to be printed is not copyrighted by a third party. Further, the customer agrees to indemnify and hold the service provider harmless for all liability, damages, and attorney fees that may be incurred in any legal action connected with copyright infringement involving the work.

Personal or Economic Rights – The customer warrants that the work does not contain anything that is libelous or scandalous, or anything that threatens anyone's right to privacy or other personal or economic rights. The provider reserves the right to use her or his sole discretion in refusing to print anything that he or she deems illegal, libelous, scandalous, improper, or that infringes upon copyright law.

Standards & Specifications

The printing industry has evolved from craft to manufacturing to service. Traditionally, the printing industry was solely a craft with very few standard operating procedures. Quality and productivity relied on individual craftsmanship; no two individuals operated on the same level. Some were very skilled and creative, and some worked faster than others. However, today, consideration of quality and productivity reigns supreme. To achieve this, the industry had to became science-oriented—with standard operating procedures that are quantifiable and measurable. Hence, standards organizations developed along with standard operating procedures.

The graphic communication industry remains an art and a science, and printing is often considered customized manufacturing because each job produced is different. However, the industry still has a long way to go to match the standard operating procedures of other industries such as auto manufacturing, aerospace, and appliance manufacturing to name a few. Over the past few decades the industry has come a long way.

According to NPES (now the Association for Print Technologies – APT), standards in the printing, publishing, and converting industry provide uniform, defined procedures and tools that help users produce quality products that are safe for their customers faster, more efficiently, and more cost-effectively. In the graphic communication industry, some of the standards and their respective organizations are:

> *Common standards and specifications arose as print became a science-based manufacturing service.*

The American National Standards Institute (ANSI) oversees the creation, promulgation and use of thousands of norms and guidelines that directly impact businesses in nearly every business sector, including printing. ANSI is also engaged in accreditation and assesses the competence of organizations in conforming to standards.

The B65 Committee, administered by NPES, develops U.S. safety standards for printing press, bindery and other printing equipment and systems used in the printing, publishing, and converting technology industry. Its subcommittees include: General Requirements, Printing Press Safety, Bindery System Safety, Bindery Cutting Machine Safety, Safety of Ink Making Equipment.

The Committee for Graphic Arts Technology Standards (CGATS) is also administered by NPES. The goal of CGATS is to have the entire scope of technical work for printing, publishing, and converting technologies represented in one national standardization and coordination effort. CGATS writes standards and provides a vehicle for other industry organizations move their work into the standards arena and approved as an ANSI standard. CGATS has standards relate to pallet loading of printed materials, terminology, plates, process control, electronic transmission of publication ads, digital data exchange, color data definition, design workflow for packaging, and ink and color characterization for packaging.

International Cooperation for the Integration of Processes in Prepress, Press, and Post Press (CIP4) is a not-for-profit international standards association with the mission of fostering the adoption of process automation in the printing industry. CIP4's membership is organizational and includes printers, prepress companies, publishers, vendors of graphic arts systems and software, integrators, distributors, consultants, and educators. It provides a variety of resources to its members and the industry, including: Software development kits, software testing systems, utilities such as CheckJDF and the JDF Editor,

International Users' Groups, Product Certification Programs, The JDF Marketplace (products and services that support JDF and PrintTalk, The JDF Integration Matrix (a reference to working integration partnerships for printers), and user case studies.

FOGRA Graphic Technology Research Association is a German-based research institute for the graphic arts. They are actively involved in maintaining several ISO standards concerning color management and printing. Based on ISO standards, FOGRA developed a system of certifications for print providers, proofing systems, and proof providers. The objective of FOGRA is to promote print engineering and its future-oriented technologies in the fields of research, development, and application, and to enable the printing industry to use the results of this activity. The association maintains its own institute, with staff members including engineers, chemists, and physicists.

General Requirements for Applications in Commercial Offset Lithography (GRACoL) was formed by the Graphic Communications Association (GCA) to develop general guidelines and recommendations that could be used as a reference source across the industry for quality color printing, focused on sheet-fed offset printing. The GRACoL Committee developed the printing guidelines that have since become de facto standards for many pressrooms. Today, GRACoL is an Idealliance Working Group, with the mission of improving communications and education in the graphic arts by maintaining the accuracy and the relevance of the GRACoL document in reporting the influence and impact of new technologies in the workflow of commercial offset lithography.

The International Color Consortium (ICC) was established by industry vendors to create, promote, and encourage the standardization of an open, vendor-neutral, cross-platform color management system architecture and components. The outcome was the development of the ICC profile specification. The intent of the ICC profile format is to provide a cross-platform device profile format to translate color data created on one device into another device's native color space. The acceptance of this format by operating system vendors allows end users to move profiles and images with embedded profiles between different operating systems. It ensures that images will retain their color fidelity when moved between systems and applications and allows a printer manufacturer to create a single profile for multiple operating systems.

The International Standards Organization (ISO) is an independent, non-governmental organization that brings together experts to share knowledge and develop voluntary, consensus-based, market relevant International standards that support innovation and provide solutions to global challenges. In printing, applying ISO standards gives a dependable result for the rate of pages yielded by a desktop printer, color reproduction when a computer file is sent to a commercial printer for printing, and even the quality of paper used for printing. There are standards for desktop printer yields for ink and color toner, for commercial color printing for matching screen color to printing color, and for paper assurances to make sure that paper is manufactured to meet the needs for different printing processes and products.

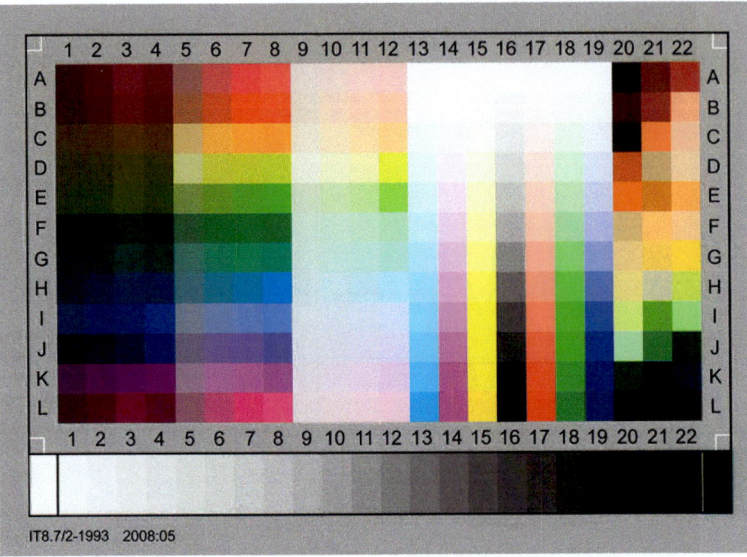

A version of the IT8 target used for scanner calibration.

IT8 is a set of ANSI standards for color communications and control specifications. Formerly governed by the IT8 Committee, IT8 activities were merged with those of the Committee for Graphics Arts Technologies Standards (CGATS). The following is a list of the IT8 standards, according to the NPES Standards Blue Book:

- Graphic technology - Prepress digital data exchange – Die-cutting data
- Graphic technology - Color transmission target for input scanner calibration
- Graphic technology - Color reflection target for input scanner calibration
- Graphic technology - Input data for characterization of 4-color process printing

Specifications for Newsprint Advertising Production (SNAP) is designed to improve reproduction quality in newsprint production and provide guidelines for the exchange of information. SNAP is intended for advertisers, advertising agencies, publishers, pre-press managers, material suppliers, and commercial and newspaper printers. Effective communication among those involved in the reproduction process ensures that the ideas of the designer and art director are printed in an accurate, efficient, and timely manner. The specifications pertain to proofing and cold-set printing by offset lithography on webs of newsprint grade paper, e.g., newspapers, pre-printed advertising inserts, and other printed materials. SNAP is not intended for magazine, catalog, packaging, or direct mail printing, nor is it intended for sheetfed, gravure, or heatset web offset processes. SNAP provides guidance for: Designers, art directors, ad agencies, editors, print buyers, marketers, color separators, merchandisers, pre-press personnel, print production personnel, electronic prepress studios, printers, media companies, and suppliers.

Specification for Web Offset Publications (SWOP) represents a set of best practices and specifications for web offset publications. They are guidelines across the industry for the preparation of print ready materials, selection of inks, protocols for proofing, and parameters for printing. While the SWOP committee worked as an independent entity in the early days, SWOP eventually joined with Idealliance. It is best known for developing scientific techniques to match color from proof to press, not only for publication (web offset) printing, but for sheet-fed offset lithography and other types of printing as well. SWOP also developed specifications for improving proofing quality and the proof to press match and certifies proofing equipment to strict color reproduction standards.

Some specific standard procedures for graphic communication are:

Computer Integrated Manufacturing (CIM) is the process of automating various functions in manufacturing and production. It involves combining hardware and software, and integrating the work through computer networks and databases. CIM is a critical element in the competitive strategy of manufacturing because it lowers costs, improves delivery times, and improves quality. It is now increasingly being used in the printing industry for producing die engravings and for die-cutting with the use of Computer Numeric Controlled (CNC) machines. Done right, it greatly reduces the number of steps necessary to produce a product.

The Job Definition Format (JDF) is a technology that allows systems from many different vendors to interoperate in automated and Management Information System (MIS) centric workflows. It is an XML software specification but also a means to allow print service providers to connect multiple vendor solutions to a MIS and to allow a workflow solution for automation and business process models. JDF is broken down into roles of MIS to prepress, finishing, digital printing, web-to-print, wide-format, etc.

The Job Messaging Format (JMF), a messaging system for JDF information, is needed in order to send information between MIS and print production systems. The format provides for a number of different types of messages, from simple notification to other devices to queries and to requests of other devices. JMF helps production workflow hardware and software systems in a JDF workflow communicate with administrative components and system controllers. The sending computer is configured to understand the receiving computer. System components can collect performance data for each piece of equipment, and information about a completed job can be sent to an accounting system for costing and billing.

12

Printing and the Digital World

Printing and the Digital World

Chapter Preview

A Brief History of Telecommunication, the Internet, and the Web

Internet-Enabled Print

The Challenge of Digital Media and the Future of Printing

SCAN

Overview

Technical innovation has always created disruptions in the *business* of print, and often does so in the *social* and *experiential* aspects of those who consume it. This was true with Gutenberg's invention as well as those of Mergenthaler, Senefelder, and Ts'ai Lun.

The rise of modern telecommunications, and especially the advent of the Internet and its visual medium, the World Wide Web, are no different. Both have disrupted the way we print and—in many cases—whether we print at all. Most of all, they have affected how people experience words and images, now mixed with such storytelling elements as video, animation, and sound.

The discussion of printed and digital media is often mistakenly framed as a zero-sum argument. Advocates of print communication too often view the Internet solely as competition for their medium. It is true that online media vies for our attention, and often disrupts the profitable status quo. However, print itself is also the beneficiary of the global connections and efficiency that the Internet provides.

When discussing how the Internet has changed print, we need to understand the "chicken-and-egg" nature of the problem. Today's information environment, especially the Web, has fostered the notion that all communication should be personalized. This puts an extreme burden on print communicators, forcing them to seek new ways to segment

Illustration ©2012 Quark Software Inc.

Introduction to Graphic Communication | 209

their message—even down to the level of one individual. However, changes in print technology itself have also altered public expectations and behavior. As Gutenberg's invention altered our views on access to knowledge, so has Internet-enabled printing changed how we view mass communication.

In many ways, the development of digital media parallels that of print, although at a much faster rate. As it was with Gutenberg, the inventors of the Web were solving immediate problems, without fully realizing the long-term consequences of their solutions. As with Gutenberg, Berners-Lee and other visionaries relied on previous technologies, using them in new and unexpected ways.

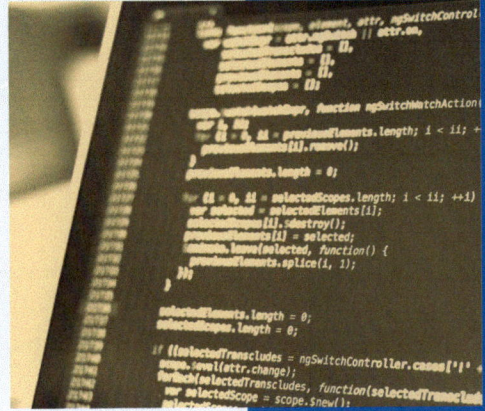

This chapter deals—all too briefly—with the history and implications of digital publishing and transactional technology as they relate to print. In many cases, online technologies have made printing far more efficient, and given designers, print buyers, publishers, and brand owners more control, and more options in a multichannel world.

Online technology has disrupted and often eliminated businesses that depended on the primacy and relative scarcity of print. However, print itself, while changed by the digital world, remains an effective and vital medium.

An analysis of print and digital reminds us that graphic communication is a single, human endeavor—not limited to the particular channels we prefer. A cave painting, an illuminated book, and an interactive app serve the same essential purpose. Only the enabling technology—the **way** we put text and images on a surface and distribute them to others—differs from one channel to another.

A Brief History

To understand how today's Internet has affected graphic communication, one must begin with electronic telecommunications. The physical network, composed of conductive wires and electromagnetic frequency transmissions, is the foundation of all digital media.

In the early 19th Century, inventors such as Pavel Shilling and Samuel Morse developed the idea that alphanumeric characters could be transmitted as electrical impulses, sent along a wire in a series of binary signals. One "language" for representing words as binary impulses still exists today—Morse Code.

Text transmission was not the only medium of telegraphy. Starting in the mid-19th Century, the transmission of images and eventually sound was also possible, using the same system of conductive wires.

Telegraphy and its successor—telephony—was an expensive proposition. The cost of creating the cables and other physical infrastructure gave rise to large, well-financed companies that dominated the new industry. These were closely allied with railroad interests having routes and stations that also became those of telegraph operations. Such companies have persisted until the present day, including global giants such as American Telephone and Telegraph Company (AT&T) and Siemens.

The network of physical cables grew rapidly throughout the 19th, 20th, and 21st Centuries, at first suspended between poles or towers but, increasingly, encased in below-ground conduits. Conductive metals such as copper are being replaced by fiber optic materials, although the process of replacement is severely limited by cost issues.

The first of many undersea cables was installed in 1850, connecting London and Paris across the English Channel. This has grown to over 700,000 miles (1.1 million kilometers) of submarine cables. Today, the Internet's reliance on undersea cables for much of its global connectivity is a significant point of vulnerability.

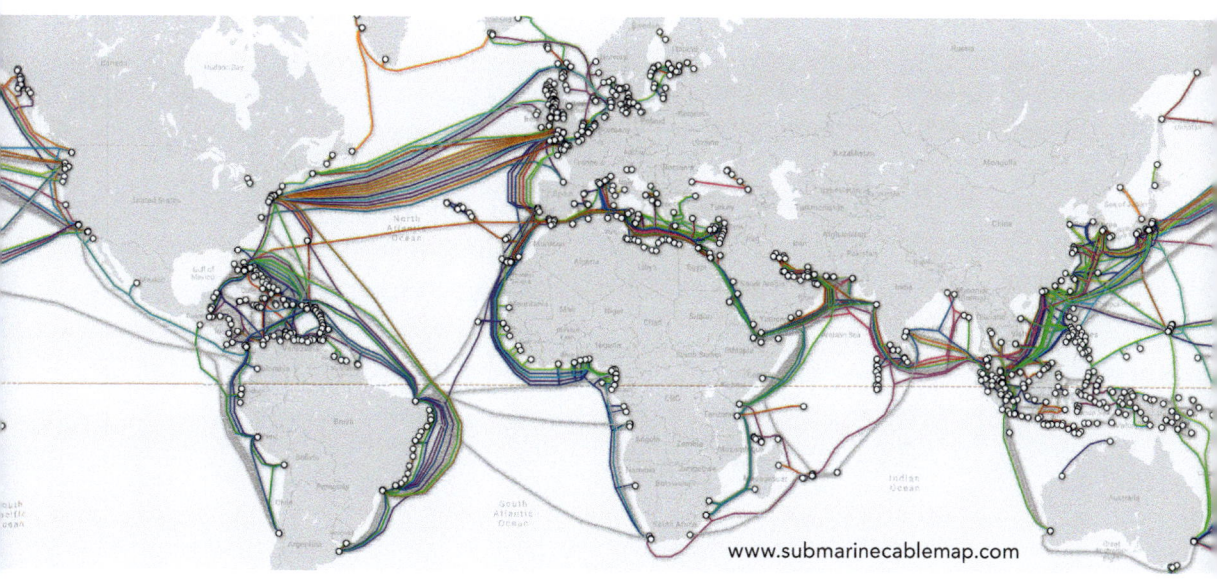

www.submarinecablemap.com

Cable Communication

The original method of Internet connectivity—using conventional phone lines and a modem—was insufficient in speed and efficiency. Dedicated T-carrier lines were considerably faster but were too costly for most applications.

Alternative telephonic protocols, such as Integrated Services Digital Network (ISDN) and, later, Digital Subscriber Line (DSL) offered some improvements. However, cable television companies now dominate the market, with their relatively recent expansion into high-bandwidth Internet connectivity.

The cable industry is dominated by relatively few, large companies—typically without regional competition. Their original purpose, beginning in the 1950s, was to provide television services to low-lying regions unable to receive broadcast signals.* Residential broadband service began in 1996, with the introduction of the "cable modem."

Cable companies' expansion from television to network services has been controversial. As was the case with their 19th Century counterparts, these private companies made large investments in infrastructure and vigorously defended their resulting monopolies. In the United States, regulators such as the Federal Communication Commission (FCC) have often been slow to recognize the significance of technological change or have been unduly influenced by powerful business interests. This was evident in the recent FCC decision to abandon "Net Neutrality," the doctrine that network service providers should not discriminate or charge differently for connectivity based on content ownership or the ability to pay more for service.

The alternative to private cable network providers is constrained by cost considerations as well as political ones. South Korea has achieved the world's most efficient Internet infrastructure using regulatory power to ensure competition and foster a rules-based approach to balancing public and private interests. The current FCC position has spurred reaction in the United States, with many calling for municipal broadband or a "nationalized Internet"—comparable to a public utility.

Broadcast and Satellite Communication

Optical fiber cables can theoretically transmit information at 99.7 percent of the speed of light and will eventually replace conductive metal as the primary network infrastructure. However, physical cables are only one "transport medium" for information. The alternative is transmission via lower-frequency electromagnetic radiation, ranging from microwaves to radio waves.

In the late 19th Century, Guglielmo Marconi proposed using radio waves as a form of wireless telegraphy. The commercial broadcast industry is based on his principle that sound can be encoded and transmitted using controlled modulation of a radio signal. The same principles were used to develop broadcast television and digital data transmission.

---

* By the 1970s, cable companies had become adversaries rather than allies of the major broadcasters. Cable companies provided their own, paid content and aggregated multiple channels into highly-profitable packages.

Ordinary radio transmission depends on a clear line-of-sight between the points of transmission and reception. Some signal frequencies are more resilient than others when it comes to signal strength over long distances. However, most ground-based broadcasting requires high towers and relays or repeaters. To overcome this inherent difficulty, broadcasters and some print publications now rely heavily on orbiting communication satellites.

(Photo courtesy of USAF.)

The first testing of communication satellites took place in the late 1950s and early 1960s. The first practical application was the launch of Syncom II in 1963 when a satellite was placed in geostationary orbit—22,300 miles above the Earth's equator—to assure constant line-of-sight with ground stations.

A geostationary orbit, first proposed by science fiction author Arthur C. Clarke in 1945, is one in which a satellite revolves at the same rate that the Earth rotates, keeping it in the same position relative to the Earth. Originally, three such satellites were needed to provide coverage for the entire surface of the planet. However, four are typically used today—each covering a smaller area and providing clearer signals and higher picture resolution at the edges of the satellite's footprint. This also allows for the use of much smaller satellite dishes, which were originally 10 to 12 feet in diameter.

Today, there are over 1,100 active communications satellites, funded by governments and private entities, plus about 2,600 that no longer work. Signals to active satellites include encoded sound, video, and data, typically using a microwave frequency as the carrier. The high cost of building and launching these satellites is magnified by the fact that they have an average lifespan of 16 years and must be periodically replaced. Some inactive satellites experience orbital decay, and burn up upon re-entry into the Earth's atmosphere, while others remain in orbit as a growing collection of "space junk."

Graphic communication—particularly daily newspapers and other time-sensitive media—frequently benefit from the immediacy of satellite communication. This was first demonstrated with Dow Jones' transmission of the entire *Wall Street Journal* to remote printing facilities, a workflow later used with *USA Today*'s full-color publication.

Not all print applications can justify the cost of satellite communication. As the bandwidth and efficiency of "ordinary" Internet connections improve, the need for line-of-sight connections will diminish.

Introduction to Graphic Communication | 213

The Internet

Regardless of the underlying telecommunications method, the Internet itself is the actual means by which graphic communication has been transformed. While a full history is outside the scope of this book, it is important to know its origins, and its similarity to the history of printing.

The Internet and the World Wide Web are terms often used interchangeably. However, they are not the same. The Internet is the underlying network structure—both hardware and software code—that connects servers and workstations and allows them to interact. The Web on the other hand provides the visual display of information (text, graphics, sound, and video) and the connections or links to other visual displays. The Web relies on the Internet to function, but they are not the same thing.

Think of the two by their analogies in the print world. A Web page is like a printed sheet of paper, while the Internet is like the presses and delivery trucks needed to create and deliver it.

The Internet originated in the 1960s, with efforts to enable computers to share information on research and development in scientific and military fields. Joseph Licklider of MIT first proposed a global network of computers in 1962, and later headed the U.S. Defense Advanced Research Projects Agency (DARPA) to develop the concept.

In 1965, Lawrence Roberts of MIT connected two computers in Massachusetts and California, using telephone lines. This showed the feasibility of wide area networking, but also revealed that

DIGITAL TIMELINE
The Internet | **The Web**

1960
DARPA Founded

1965

ARPANET Launched — Compu-Serv*
1970 — Project Gutenberg (first eBook)*
TCP/IP & FTP Protocols — Email*

Altair
1975
CERNET

First Commercial Cellular Network
1980
BITNET (Because Its Time Network) at IBM — "Smart" Modems — WHOIS Search Engine*

Macintosh
Domain Name System (DNS) — **1985**

ISDN — Web protocols (HTML, HTTP, URLs)
1990
WaveLAN (WiFi precursor) — First Web Page — Gopher
PDF* — Delphi — Mosaic**
Navigator**
1995 — Internet Explorer** — CSS — Opera**
JavaScript
DSL — First Weblog (blog)

2000 — MOBI (eBook format)*
Kindle
Safari**
FireFox** — First Podcast
2005 — AJAX
iPhone — EPUB (eBook format)
XML 1.0 — Chrome**
2010 *iPad* — EPUB2
CSS3
HTML5
2015 — Microsoft Edge**
EPUB3

* = non-Web visual/communication milestone
** = Web Browser Software
Dates are approximate. ©2018 Harvey Levenson & John Parsons

telephone circuit switching was inadequate to the task. In the meantime, however, Leonard Kleinrock of MIT, and later UCLA, had developed a theory of "packet switching," which would eventually form the basis of Internet connections. Roberts transferred to DARPA in 1966 and began development of the Advanced Research Projects Agency Network, or ARPANET.

Launched in 1969, ARPANET originally connected computers at four major universities, UCLA, Stanford, UC Santa Barbara, and the University of Utah. More computers were added over time, eventually transforming the wide area network into what we know as the Internet.

In 1972, open architecture components were added to ARPANET, including what became known as the Transmission Control Protocol (TCP) and the Internet Protocol (IP). TCP addresses reliability issues, such as the order in which data arrives and the handling of errors and traffic control. IP deals with identifying nodes on the network and routing packets to the correct location.

By using TCP/IP and other protocols, the Internet allows us to enter, manipulate, and view data from any authorized system, wherever that data may reside.

The World Wide Web

Early users of the Internet, particularly scholars and librarians, did not have a particularly easy task. There was no user-friendly interface, and early workstations operated on a time-share basis, meaning that results could lag far behind initial queries. TCP/IP was unavailable on early personal computers, where network connection (if any) was limited to slow Local Area Networks or LANs.

Sir Timothy John Berners-Lee, inventor of the World Wide Web

In 1989, Oxford Professor Timothy Berners-Lee proposed the idea that data encoded in "Hypertext"—more specifically the Hypertext Transfer Protocol or HTTP—could be communicated between a server and a client computer via the Internet. That same year, he successfully demonstrated his theory.

Hypertext was a concept championed by visionaries such as Vannevar Bush, Douglas Engelbart, and Ted Nelson. Briefly, it involves text on a computer display that contains links (called hyperlinks) to other documents, allowing the user to immediately access additional and presumably relevant information.

Berners-Lee's implementation made hypertext a reality. By combining hypertext with TCP and the Domain Name System (DNS) for identifying specific network entities, he created the World Wide Web. On the very first Web page, created at CERN (the European

Organization for Nuclear Research), he defined it as "a wide-area hypermedia information retrieval initiative aiming to give universal access to a large universe of documents."

The first implementation of HTTP was not the user-friendly experience associated with the Web today. However, in 1993, Marc Andreessen and his team at the National Center for Supercomputing Applications (NCSA) developed the first graphical browser for the Web: Mosaic. Later, Andreessen founded Netscape, which developed Navigator, the first commercial Web browser.

All browsers use the Hypertext Markup Language (HTML) to describe the structure of a Web page and Cascading Style Sheets (CSS) to control its presentation. HTML and CSS have undergone many changes, to facilitate the evolving nature of the Web, and are currently maintained by the World Wide Web Consortium (W3C), directed by Berners-Lee.

Other Internet-based services, including email and the File Transfer Protocol (FTP) would also complement the Web, and transform the way we work. However, it is the World Wide Web that has made printed graphic communication more complicated and—at the same time—a more powerful, versatile medium.

Electronic Paper and Mobile Displays

One other historic development warrants discussion when considering the impact of digital technology on print: electronic paper or e-paper. Unlike its fiber-based namesake, it is not a consumable component, but a viewing technology created to emulate the qualities of a conventional reading surface—paper.

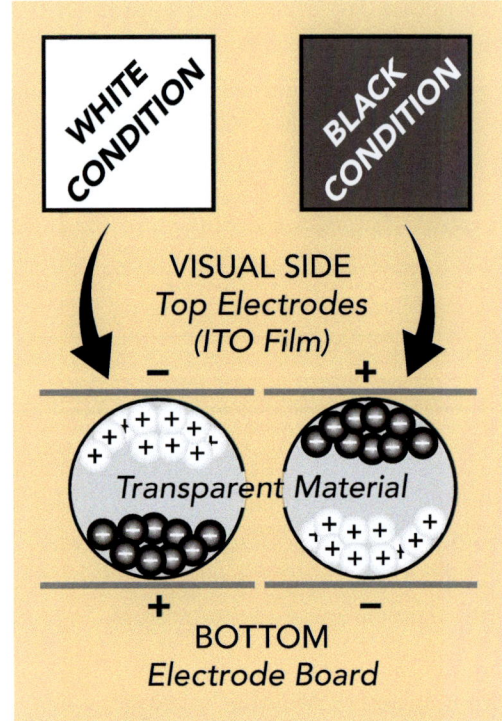

In electronic paper, based on the MIT model, positively-charged white particles and negatively charged black particles are attracted to their oppositely-charged electrodes. (Image courtesy of Midori Mark.)

First developed at Xerox PARC in the 1970s, e-paper was an alternative to backlit displays such as cathode ray tubes (CRTs), as well as displays based on light-emitting diode (LED) or liquid-crystal display (LCD) technology.

Unlike conventional displays, electronic paper reflects ambient light, rather than transmitting light to the human eye. (The color implications of this distinction were discussed in Chapter 6.) This would make it more comfortable to read and provide a wider viewing angle.

In the 1990s, researchers at MIT's Media Lab developed a less costly version of e-paper, based on small, charged particles suspended in a transparent oil, and manipulated to either a black or white condition based on charged electrodes. The same researchers later founded the E Ink Corporation, which produced the display technology used in eBook readers such as the Amazon Kindle.

The attraction of electronic paper is in its potential to make content consumption more efficient. After reading an e-paper version of Gone With the Wind, one would be able to replace it instantly with From Here to Eternity, without having to shelve and manage a growing volume of paper. This has proven to be true—but only to a degree—in book publishing.

Although color versions are under development, it is unlikely that electronic paper will displace conventional paper in most applications. The primary obstacle is cost, as Hearst discovered when it produced a hybrid paper/e-paper cover for Esquire magazine.

Mobile device manufacturers—the most logical users of e-paper technology—have focused far more on touch-screen LCD displays. So, although electronic paper remains viable, it is mobile technology itself that poses the most significant challenge to the role of print.

The rise of mobile, combining Internet connectivity, the Web, and device features such as cameras and touch-responsive screens have led to a different type of publishing medium—a hybrid of print and online experience. This book, for example, employs Ricoh's "Clickable Paper" technology to combine the two learning experiences.

In 2008, Hearst Corporation published a newsstand-only edition of Esquire with "the world's first e-ink cover." Powered by two batteries hidden inside the front cover construction, the small black-and-white display featured five words of text that flashed on and off next to pre-printed color elements.

Internet-Enabled Print

Graphic communication is driven by the need to tell a story or deliver an important message. Such expression evolved from simple markings on stone and clay to the mass production of printed paper, as described in Chapter 2. From Gutenberg to today, each new innovation has increased the volume, efficiency, and accessibility of print.

Print's ubiquity fostered the spread of new ideas—good and bad—and made literacy a more essential human capability. It created a cultural mindset, based on interpretation of static words and images. The introduction of sound and motion did not eliminate reading, but rather added new media "channels"—namely radio, film, and television.

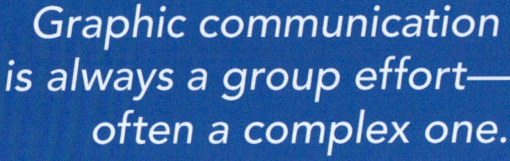

Graphic communication is always a group effort— often a complex one.

Likewise, the addition of the Web and other online media to an already crowded communication environment did not eliminate print or other "legacy" media. However, it created new expectations about how we generate, manage, and consume the printed word.

Graphic communication is always a group effort—often a complex one. Despite the inherently self-contained and empowering nature of personal computers, designers and print professionals inevitably must collaborate. To do so, they must connect their computers with each other, and with a complex array of file servers, workflow systems, and output devices.

There are many ways in which the Internet can do exactly that. One is the actual creation and automation of printed pieces. Another is the process of print procurement, known as Web-to-Print (WtP). A third involves overall management, not just of printable products, but of their underlying data and the ways they can be repurposed as well.

Network Evolution

Collaboration depends on a network infrastructure that is both robust and secure. This was not the case in the early days of digital publishing and is still not optimal outside developed countries and major metropolitan areas. However, as high-bandwidth connections become more prevalent—and safer—the creation and consumption of print will invariably become more efficient.

Graphic communication workflows come with heavy network requirements. Early Local Area Networks (LANs) and dial-up Internet connections were no match for the large file sizes needed to produce a page. Early DTP designers and service providers had to rely on removable hard drives to manually transfer large image and PostScript files—a process jokingly known as "sneakernet."

Today, Internet bandwidth speeds have improved to the point where some services can be handled at the network level. File transfer is one of these. However, CPU-intensive tasks, such as image manipulation and page layout, are still done mostly on individual computers, not in the cloud. (Adobe's Creative Cloud is not a true, client-server cloud service, but a *subscription and authentication* model for installed, desktop software.)

Despite the inherently personal computer-centric nature of print design and production, an increasing number of collaborative, print-related services are in the cloud. They use the Software-as-a Service (SaaS) model—where the actual data processing is done on Internet servers and databases, controlled remotely at the browser or app level.

Many of these SaaS systems began as proprietary software, or as client-server systems on Ethernet LANs, but have migrated to SaaS as the Internet has evolved. All support the notion that graphic communication is a group effort.

Document Composition

Computerized page layout began well before network-enabled cooperation was possible. In the 1960s, technology pioneer John W. Seybold realized that computers could control every aspect of preparing, manipulating, and formatting written text. However, this required sophisticated software and hardware to do so. It also required a **page description language** to turn computer data into human-readable form.

Using a digital language to define a page predates PDF by many years. Donald Knuth's TEX (pronounced 'tɛk'), still considered one of the world's most sophisticated typographic languages, was released in 1978. Many other languages were developed—usually in proprietary formats—before Adobe's Xerox PARC-inspired PostScript (and later PDF) became the *de facto* print standard.

John W. Seybold and his son, Jonathan, developed and reported on computer-based composition and publishing systems—later including DTP and the Web—for many years.

These languages—and the systems that use them—were the precursors of modern, digital page design, including Aldus PageMaker, QuarkXPress, and Adobe InDesign. They also made possible another, less glamorous trend: database publishing.

Before Web-based collaboration was possible, early database publishing systems, such as 3B2 and Xyvision, could create more pages at far less cost than could design-oriented systems. They could automatically insert complex data into multi-page documents, leveraging the inherent structure of the page description language. The need for accuracy and production speed far outweighed aesthetic design concerns.

This process and the systems that enabled it were generally reserved for large companies with sophisticated networks and database infrastructure. However, the model of using organized data to produce structured documents became more widespread with the Web, and with SaaS-based workflows.

To succeed, however, Web-based document composition needed a reliable template model, and a robust method of editing and customizing online documents for printing.

Templates, Versions, and Online Editing

Print buyers frequently have two, often conflicting needs. Creating a unique, compelling design is essential for a company's public brand image. This is the "sweet spot" for page layout software. However, the same companies also need multiple versions of the same piece. These can be just a few, regional variants, or millions of personalized versions, as described in the Variable Data Printing (VDP) portion of Chapter 8.

Template-based workflows are at the heart of many Web-to-Print (WtP) systems for ordering printed materials. Users simply enter data into designated fields, and the system would apply that data to an exact position in online template—preferably one with uniform dimensions and element positions.

The process can also be more automatic and on a larger scale. Instead of manually entering data in fields, the print buyer may select an Excel CSV or other data file

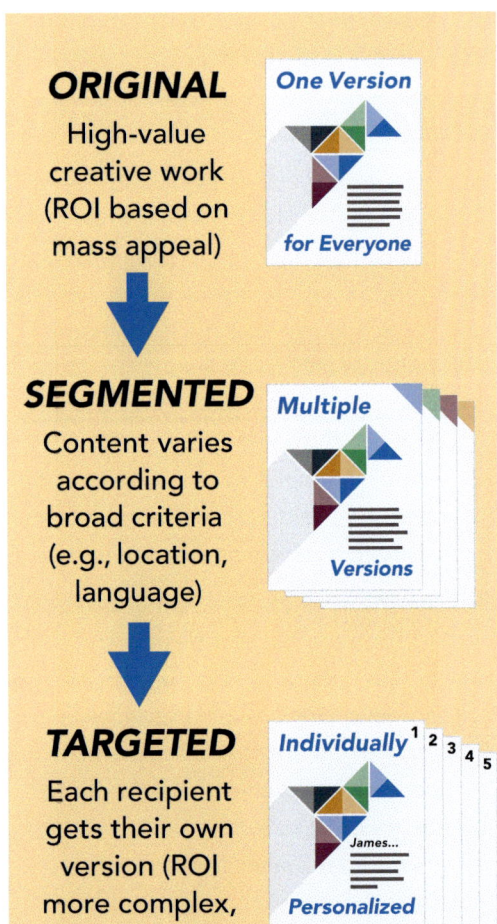

The use of templates for printing requires a robust page composition architecture, reliable data-handling, and an efficient Internet infrastructure.

to create multiple, individualized versions of the printed piece.

Although print buyers are usually familiar with a template's design, they still need to see a proof of the customized version before giving approval to print. This is most often provided in PDF format. On-screen renderings in Adobe Flash or HTML5 are also possible. Once approved, a PDF production file is generated, sent to the prepress workflow, and printed with little or no manual intervention.

This approach must balance potentially competing interests. The brand owner will always require that certain graphic elements, such as logos, typefaces, and corporate colors, be "locked" in the approved design of a page. However, those seeking to create a custom version of the document may have valid reasons to make changes not anticipated by the original designer. Most template-based WtP systems employ a sophisticated users and permissions structure to create and manage the rules of engagement.

The templates themselves are typically created in page layout applications—mainly Adobe InDesign. Early WtP systems relied heavily on proprietary technology to merge customer-supplied data with custom-programmed print templates. However, with the release of Adobe InDesign Server in 2005, and the introduction of the InDesign Markup Language (IDML) in 2008, the process became more predictable and less costly.

For business printing—such as brochures, data sheets, advertisements, and business cards—template-based WtP systems are usually adequate. However, for many applications, there is a need for greater creative freedom. Users that do not need or cannot afford the complexity of InDesign often want document editing tools—preferably online and as part of a SaaS environment.

Early WtP systems often employed Adobe Flash to simulate the printed piece on screen and provide basic editing capabilities such as font and color selection and the ability to move or scale

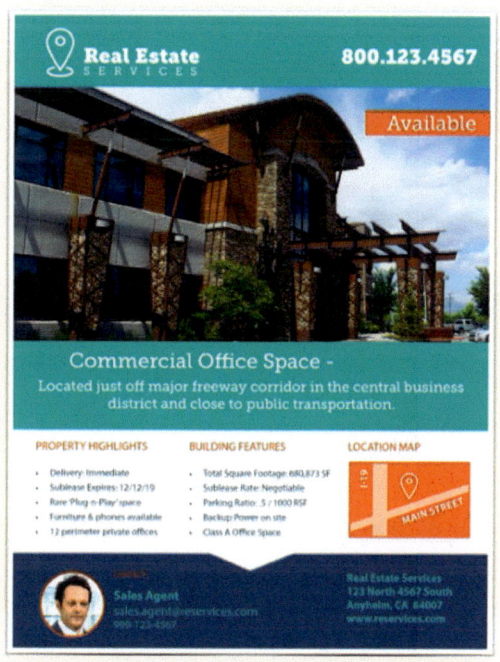

Online, SaaS-based editing tools typicaly emulate those of page layout software. In the left-hand view, brand-critical elements, shaded red, are "locked." Editable images and text, shaded blue and green, respectively, can be modified by authorized users. Lest dollicilita ea voluptatibea cume qui aut alitat ipit facea vel et quat minusdanda cus pero estia vernatius. (Image courtesy of Silicon Publishing.)

page elements. Although this provided some level of WYSIWYG ("What You See Is What You Get") interactivity, it lacked the precision of true page layout software. Flash also worked on only one "page" at a time, and often fell short with regard to font handling and other norms of desktop publishing.

With the rise of mobile devices and the corresponding unpopularity of Flash, finding an alternate approach to online editing became imperative.

The challenge of online editing stems from the sheer complexity of page layout software itself. It must perform far more tasks than most other applications—from typographic control to page geometry and image placement, all with continuous screen rendering to reflect each decision of the designer.

On increasingly powerful desktop computers, these tasks are handled well by Adobe InDesign and QuarkXPress. However, the bandwidth limitations of Internet connections have until recently made a "client- server" approach to page layout problematic.

In 1999, Adobe released InCopy, a text editing version of InDesign. It allowed users to "check out" and edit stories within an InDesign layout, and even facilitated an online editorial workflow. However, InCopy was a desktop application, not a SaaS-based editing tool. The real breakthrough came in 2005, with the release of InDesign Server (IDS).

Initially, IDS was used primarily to facilitate WtP systems, allowing template-based documents to render accurately, both on screen and in print. Eventually, however, third-party developers began to use IDS to develop online editing tools comparable to Adobe InDesign. Early attempts to emulate desktop layout tools using Flash have now been largely replaced by HTML5.

The Commerce Aspect of Web-to-Print

During the dot-com boom of the late 1990s, over 140 companies theorized that print job ordering, collaboration, customization, and even design could be done more efficiently on the Web. Most of those companies disappeared, were acquired, or changed business models during the 2000-2001 downturn. However, their core idea survived. Today, the Web-to-Print process has several main components:

- **Online ordering,** typically through a Web "storefront" of commonly printed products.
- **Customization** of a single print job or a batch of similar jobs. (This is arguably the most significant aspect of WtP, giving designers the ability to edit online.)
- **Collaboration,** allowing print buyers, designers, and production professionals to communicate online, using a visual representation of the printed piece.
- **Back-office integration,** connecting online ordering and fulfillment to existing enterprise systems.

Most WtP storefronts today are run by individual printing companies, as an adjunct to their main Web presence. (A few from the early dot-com era still exist, serving as sites for brokering print jobs to

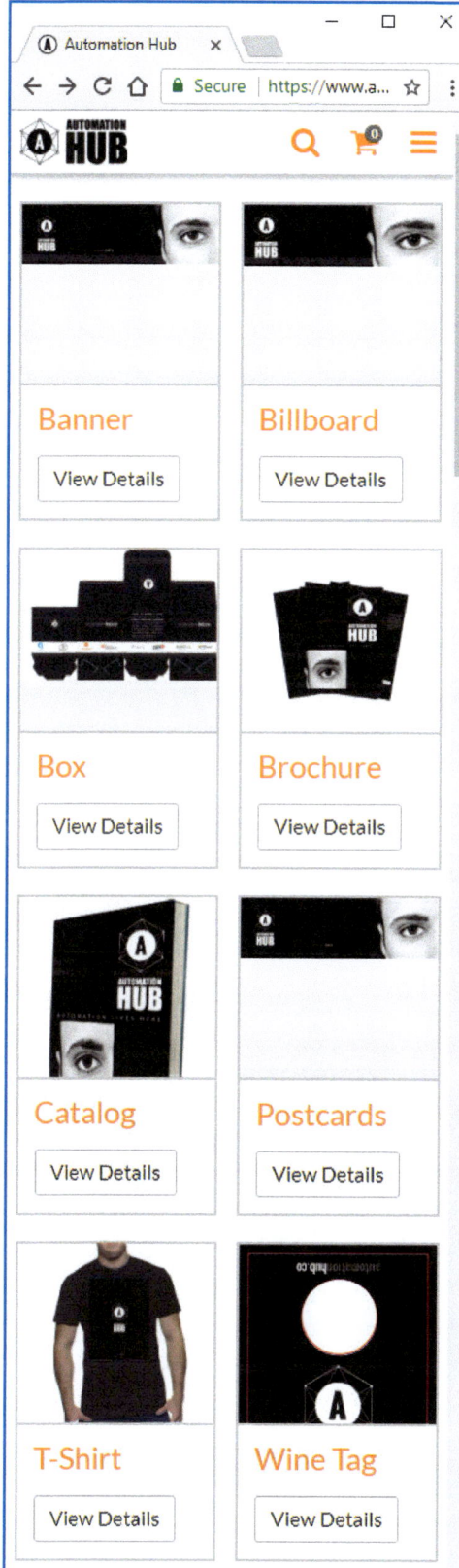

the lowest bidder.) A printer's WtP site can either be focused on individual customers (Business-to-Consumer or B2C) or on companies that need printed collateral (Business-to-Business or B2B).

B2C Web storefronts often feature small, one-time print jobs typically ordered from a commercial or quick printer. These can include greeting cards, photo albums, event announcements, programs, and business cards. By offering their services online, printers can extend their geographic reach to almost any location. However, by doing so, printers are also adding to the number of competitors in that location, making it easier for customers to demand lower prices.

Consumer WtP storefronts emphasize individual personalization, as covered later. In contrast, B2B Web storefronts emphasize the routine ordering of materials such as product brochures, sell sheets, forms, signs, training materials, and other printed collateral. Typically produced on digital presses and in pre-determined sizes, such online ordering can keep inventory levels low, or even non-existent—producing materials on demand or "just in time" to meet the immediate need.

Both types of WtP storefronts require adequate security and control of user permissions—particularly regarding payment or authorization. B2B storefronts may also be connected to Enterprise Resource Procurement (ERP) systems on the client side, and to inventory systems on the printer side.

Web-to-Print storefronts often include a catalog of available products, ways to customize each job, and a means of specifying product type, quantities, sizes, and paper choices. (Image courtesy of Aleyant.)

Managing Digital Assets

Print design involves a large number of visual elements, each one subject to its own revision cycle. The potential for chaos increases when multiple decision makers are interacting across the Internet, and when one considers the many versions and variations of each printed piece using these images.

> **The cost of searching for the right images can be 20-30 percent of a worker's time.**

The cost of searching for the right images or other assets is high—between 20 and 30 percent of a worker's time, according to some estimates. This does not count the time spent re-creating lost images.

So, to reduce those costs, companies of all sizes increasingly resort to (DAM) systems to organize and manage images for both print and digital media.

A simple image database such as Adobe Bridge can serve as a "poor person's DAM" for individual designers. However, it is not URL-based, and cannot meet the needs of larger organizations. A true DAM must do more, including user access control (who checked out what asset), version tracking, and integration with other, server-based systems.

DAM was originally an on-premises system, running on dedicated servers and connected to personal computers via an Ethernet LAN. This model is being supplanted by the cloud. Most major DAM system vendors are offering SaaS versions of their product, moving from the local network to a much larger one—the Internet itself.

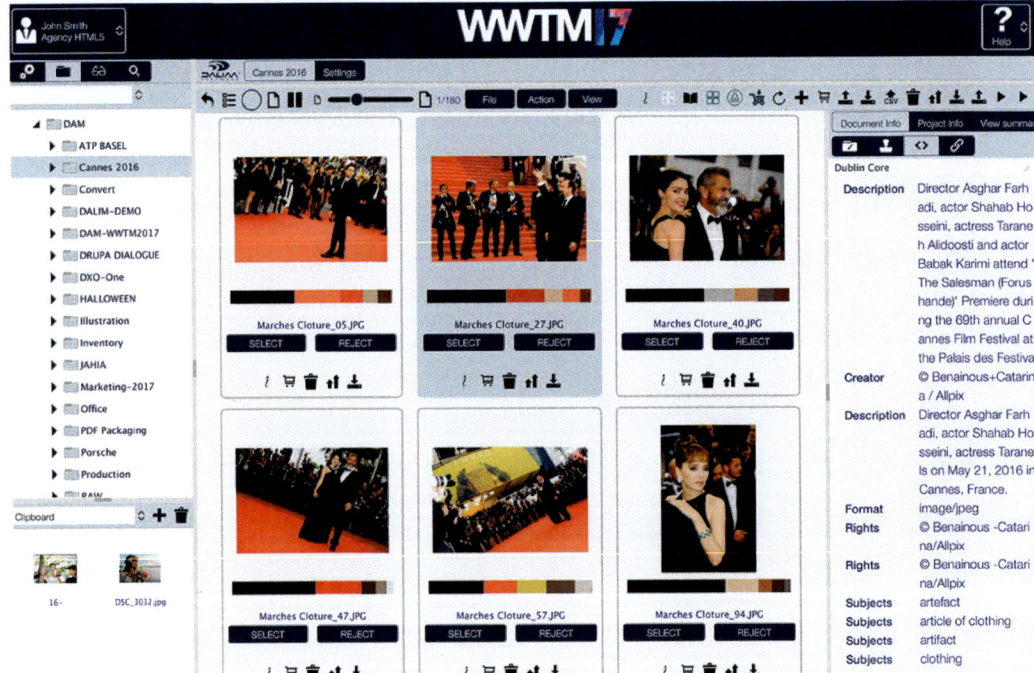

A high-end DAM, showing the metadata, right, for each image. (Image courtesy of Dalim Software.)

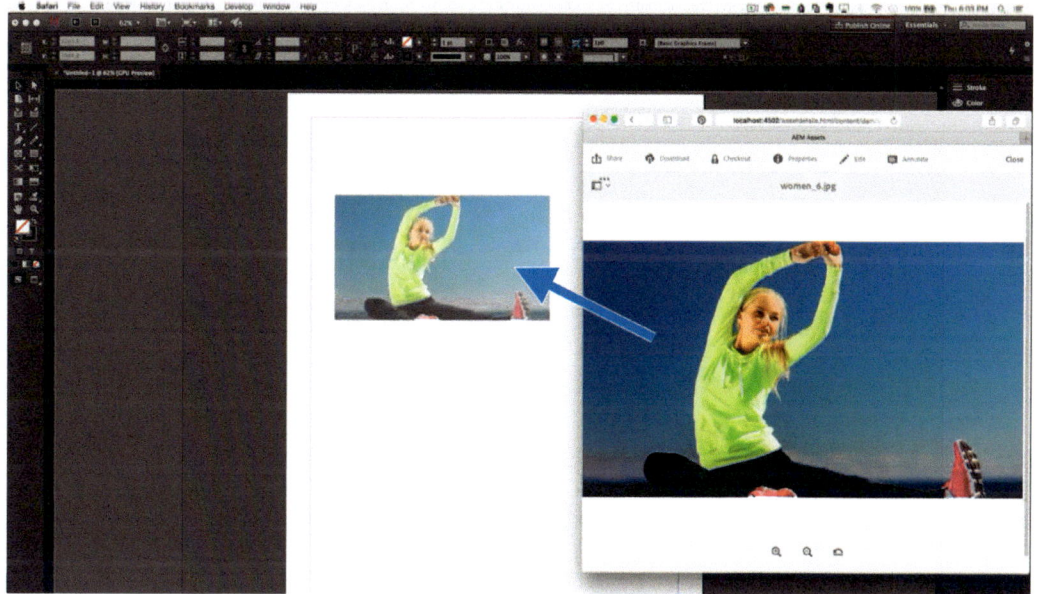

With an InDesign plugin, designers can directly access a DAM repository, dragging and dropping the right image into the layout, inserting a URL-based link. Subsequent changes made in the DAM can be automatically updated in the layout. (Image courtesy of Silicon Publishing.)

Asset management systems store not only the image or file itself but also the descriptive information or **metadata** about that resource. This can be quite extensive, including subject categories and descriptive text (for search purposes), creation and modification dates, authorship, and licensing rights. An image may be stored multiple times, in different revision stages, resolutions, and file types. These variations are usually stored under a common identification number.

It is difficult to enter and maintain DAM metadata consistently. Fortunately, the process has improved somewhat, as systems become more automated and users develop new digital work habits.

A designer's ability to find the right asset is only one benefit of DAM. A "drag-and-drop" connection to an image repository uses the designer's normal process, while also creating a "live link" to the managed asset. If the linked image is revised or updated, the layout can automatically reflect a decision made at the DAM level.

Asset management is not limited to print. Websites and mobile apps also benefit from a well- structured, easily accessible repository of images. Increasingly, publishers and marketers are using DAM to produce effective graphic communication for multiple output channels.

Finally, DAM relies on the ability of designers to create and upload successive versions of their images, usually from Adobe Photoshop or Lightroom. This gives the designer or photographer the responsibility to create as much metadata about the image as possible, using the "File Info" features of those programs. Most DAM systems can read and recognize this information. However, except for automatically-generated metadata such as creation and modification dates, the lack of data entry will make the DAM system less valuable.

The Challenge of Digital Media

Despite the benefits of Web-based technology to print workflows, many still consider digital media as an existential threat to the printing industry.

This fear is not entirely unfounded. The efficiencies and globalization potential of Internet-enabled printing will reduce the number of traditional jobs and increase offshore competition. Some formerly-profitable sectors, such as newspaper coupon inserts and direct mail, will see significant losses to their mobile and online counterparts.

Such disruption is not new—and not a valid reason to view print and digital as a zero-sum problem. Print will be forever changed by the digital world, but it will not be replaced by it.

The Haptic Advantage

As discussed in the Introduction, print provides a multi-sensory experience that screen-based media cannot. It is both visual and tactile. Research in the field of haptics (the science of touch) has found that comprehension, long-term retention, and perception of value are significantly higher with print than they are with online media—even for "digital natives."

This means that preference for paper is not simply generational, or something merely familiar or comfortable.

This is borne out in the recent trends involving eBooks. According to the Book Industry Study Group (BISG), the percent of book buyers who read eBooks daily or weekly rose sharply from 2010 to 2012, from 5 percent to over 20 percent, but has remained the same or declined since then. The purchase of eBooks overall has also declined precipitously during the past two years.

This does not diminish the intrinsic value of digital media. Unlike print, online media can be updated at any time. It can also provide immediate "findability" and access to related materials. Unless it is connected to online information (as this book is), print cannot provide that level of flexibility or real-time interactivity.

Haptics (touch) is a human experience that digital technology cannot easily replicate.

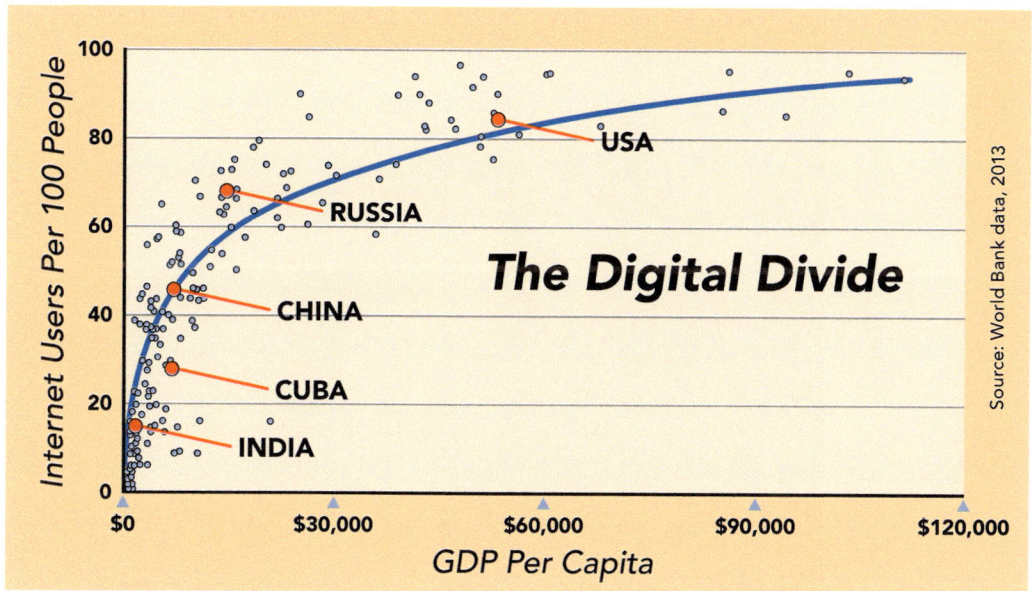

Digital Discrepancies

Print and digital media have another important difference: economics. While it is popular to assume that the Internet is everywhere, access is still limited by geography and particularly by the affordability of devices, content and service subscriptions, basic connectivity, and even electrical power. Mobile Internet is starting to make a difference in countries such as India. However, printed media have no such hidden costs.

With print, all of the energy, creativity, and investment involved in its creation occur before the finished product ends up in the user's hands. Except for the need to store and ultimately dispose of it, a printed piece requires only a user's eyesight and the time to read it. Even its legal status is a settled issue. One may not make illicit copies of a printed piece, but the physical copy is legal property—to sell, loan, or bequeath to one's heirs.

Digital media are more complicated. Connectivity, a device, and a level of technical expertise are needed to experience an online site, an app, or a piece of digital content. They also have decidedly different legal status. Instead of owning a printed copy, the user must license it, as is the case with a software application. The actual owners are often large corporations that maintain strict control and do not guarantee that they (or the content) will exist in the future.

As "containers" for information and narrative, both print and digital provide distinctly different value, and impose very different requirements on the end user. In many cases, it is not a question of using print *or* digital, but of using both.

The Future of Print

The ultimate challenge for graphic communicators is not to defend print against the encroachments of digital, but to discover ways in which the two can be used together. This book is one such attempt, providing the haptic experience of a printed book and the means to access related, online information and interaction. Other solutions are possible.

Introduction to Graphic Communication | 227

One of these is the idea of **repurposing**. Content can and should be available in more than one format, provided that each format has its own, intrinsic value. Simply copying content from a print layout to a webpage is insufficient.

The notion of repurposing is at the heart of **multichannel publishing**, the use of every possible means to convey text and images to the reader in their preferred medium. Content Management Systems (CMS) offer solutions, primarily for the Web but also for Web-and-print applications. Formerly print-centric developers are exploring new ways to meet this need.

Page layout vendor Quark Software is one example. Their App Studio system begins with page layout files—from QuarkXPress or InDesign—or with an XML data repository. From the extracted data, which is not constrained by a print page layout, is converted to HTML5 and then "wrapped" in a responsive app for mobile devices of any screen size.

Adobe and other developers have made similar moves. Adobe's Digital Publishing Suite (DPS), now part of Adobe Experience Manager, allowed users to generate interactive apps from InDesign layouts. The page-centric nature of InDesign made the original DPS problematic for multiple screen sizes, however.

Many others are attempting to solve the multichannel challenge, including developer mag+, whose cloud-based service generates interactive, responsive apps from InDesign layouts, using a plugin.

Print is by no means at risk from the encroachments of digital. While aspects of the printing business will be impacted—even fatally—print itself will be strengthened and changed for the better. As we approach the 30th anniversary of the Web, print will emerge as a coequal, and in some cases superior medium among many others.

The human practice of telling stories and seeking new audiences for them will not abate—nor will it be confined to established ways to create "story containers." There will always be a new way to use words, images, and actions to communicate meaning. Finding out how to do that well is the ultimate opportunity.

Epilogue, Bibliography & Index

Epilogue

Overview

Beginning the Next Conversation

A Starting Point for Everyday Resources

User-Generated Video and Other Content

Interactive Media: www.**igcvideo**/Epilogue

Web Links: www.**igcbook**/Epilogue

The Conversation Is Just Beginning

When we first set out to revise a 10-year-old textbook on graphic communication, we did not fully realize what we had started. What began as a means of avoiding clunky QR Codes grew into what we suspect may be a new way of fusing the print and digital learning experience. We learned a lot in the process, as we hope you will. However, we do not know what will happen next—with graphic communication in general or with this new approach. Here are some ideas:

First, as we noted in the Introduction, this book will serve as a starting point for exploring online resources related to graphic communication. Besides the videos themselves, there are many helpful calculators, data, lists of companies and equipment, and other valuable online resources that will make your work easier. Simply scan this spread with the Clickable Paper app to access an ever-growing list.

Graphic communication answers are only a scan away...

Resources available by scanning this spread include Web-based tools, calculators, and data tables for graphic communication professionals.

If your favorite calculator for paper specs, color values, binding styles, or other data are not there yet, please let us know: book@intuideas.com

Introduction to Graphic Communication | 231

Of course you can also find and save such resources on your own, but we believe that a book, a physical object with intrinsically valuable content, will serve as a better starting point for finding what you need.

The second reason for this epilogue is a request for online content from you.

If you scan the Epilogue's opening spread around the time we publish (April 2018), you will find a few videos, but not as many as with other chapters. Over time, we hope that number will grow. We hope the online learning portal for the Epilogue will become a place for new videos and interactive material—yours.

If you or your class or company want to explore topics on graphic communication—and record a video in the process—this will be a place for it. It can on a topic stemming from the book, or something we never even imagined. Production quality is not an issue. A recorded Google Hangout or a full-blown documentary are equally acceptable. The submission doesn't even need to be a video.

Whatever you decide to do, let us know and we will do all we can to add your contribution to the ongoing discussion. You can use that quaint, nearly 50-year-old medium, email: **book@intuideas.com**.

Graphic communication is all about having a conversation. Let's start one.

— Harvey Levenson & John Parsons

Bibliography

Adams, Michael J., David D. Faux, and Lloyd J. Rieber. *Printing Technology*. Albany, New York: Delmar Publishers, Inc., 1988, 629 pp.

Agfa. *Digital Color Prepress* (four volumes) and *Digital Photo Imaging*. 1992-1997.

Alexander, George. "Who Are the Digital Printers?" *Digital Publishing Solutions*, March 2005.

Alexander, George, Stephen Edwards, and Kurt Wolf. "'Gang of Four' Introduces New Job Ticket Format." *The Seybold Report on Publishing Systems*. Media, Pennsylvania: Seybold Publications, April 3, 2000.

Allenby, Braden R. and Deanna J. Rickards (eds.). *The Greening of Industrial Ecosystems*. Washington, D.C.: National Academy, 1994.

Anderson, R. *Mid-Course Correction: Towards a Sustainable Enterprise: The Interface Model*. Atlanta: The Peregrinzilla Press, 1998.

Anon. *British Printer*, October 1998, www.uidaho.edu/pd/imgsett.htm

Anon. *Digital Basics*. Cohoes, New York: Mohawk Paper Mills, 2001, 56 pp.

Anon. *Paper Knowledge*. Chillicothe, Ohio: The Mead Corporation, 1990, 474 pp.

Anon. *Print 2000*. Arlington, Virginia: Printing Industries of America, Inc., 1990, pp. x-1–x-5.

Anon. "The Digital Press Revolution, Digital Presses: Now a Staple in the Graphics World," *Digital Output*, April 2003.

Anon. *The Printing Service Specialist's Handbook and Reference Guide*. Alexandria, Virginia: Society for Service Professionals in Printing, 1994, pp. G.1–G.37.

Anon. "What Is a Page Description Language (PDL)?" www.cs.wpi.edu/~kal/elecdoc/EDpdl-def.html

Apps, E.A. *Ink Technology for Printers and Students*, Vol. I-III. New York: Chemical Publishing, Co., Inc., 1964, 256, 347, and 293 pp.

Arnold, Edmund C., *Ink on Paper: A Handbook of the Graphic Arts*. New York, Evanston, San Francisco, London: Harper & Row, 1972, 374 pp.

Avery, Allen. "Remote Proofing: Close at Hand." *American Printer*, February 1, 2002.

Baird, Russel N., Duncan McDonald, Ronald H. Pittman, and Arthur T. Turnbull. *The Graphics of Communication*. New York, etc.: Harcourt Brace Jovanovich College Publishers, 1993, 410 pp.

Bann, David, and John Gargan. *How to Check and Correct Color Proofs*. Cincinnati: North Light Books, 1990, 143 pp.

Beach, Mark. *Graphically Speaking, An Illustrated Guide to the Working Language of Design and Printing*. Manzanita, Oregon: Coast to Coast Books, 1992, 321 pp.

Beach, Mark., Steve Shepro, and Ken Russon. *Getting It Printed*. Portland, Oregon: Coast to Coast Books, 1986, 236 pp.

Bear, Jacci Howard. Your Guide to Desktop Publishing. http://desktoppub.about.com/

Beck, Ulrich. Risk Society: Towards a New Modernity. Newbury Park, California: Sage. 1992 [1986].

Benyus, Janine M. *Biomimicry: Innovation Inspired by Nature*. William Morrow & Co., 1997.

Bisset, D. E., and C. Goodacre, H. A. Idle, R. E. Leach, and C. H. Williams. T*he Printing Ink Manual, 3rd Ed*. United Kingdom: Van Nostrand Reinhold Co., Ltd., 1979, 488 pp.

Bixler, R., and M. Floyd. "Nature Is Scary, Disgusting, and Uncomfortable." *Environment and Behavior*, 5(2), 1997, pp. 202–247.

Blair, Raymond N. and Charles Shapiro (eds.). *The Lithographers Manual*. Pittsburgh: Graphic Arts Technical Foundation, 1980.

Blatner, David. *Real World Scanning and Halftones*. Berkely, California: Peachpit Press, 1998, 464 pp.

Bridg's. *CTP Handbook for the Graphic Arts*. South Holland, Illinois: IPA, 2000, 24 pp.

Broekhuizen, Richard J. *Graphic Communications*. McKnight Career Pub., 1973, 380 pp.

Brown, L., M. Renner, and C. Flavin. *Vital Signs 1997–1998: The Environmental Trends Shaping Our Future.* London: Earthscan, 1997.

Brown, Lawrence D., and Marcus L. Caylor. *The Correlation between Corporate Governance and Company Performance* (white paper). Institutional Shareholder Services, 2004.

Brundtland, G-H. (chair). *Our Common Future: Report of the World Commission on Environment and Development.* Oxford: Oxford University Press, 1987.

Bruno, Michael H. (ed.). *Pocket Pal.* Nashville: International Paper Co., all editions.

Bruno, Michael H. *Principles of Color Proofing: A Manual of the Measurement and Control of Tone and Color Reproduction.* Salem, New Hampshire: GAMA Communications, 1986, 395 pp.

Butz, Christopher, and Andreas Plattner. *Socially Responsible Investment: A Statistical Analysis of Returns.* Basel, Switzerland: Sarasin Sustainable Investment, January 2000.

Carruthers, Roderick W. *Why Won't You Buy Your Printing from Us? A Study of Outside Print Buyer Attitudes.* St. Albans, Vermont: Strategic Management Consulting, Inc., March 1984 (prepared for Printing Industries of Northern California).

Cavuoto, James, and Stephen Beale. *Linotronic Imaging Handbook.* Torrance, California: Micro Publishing Press, 1990, 218 pp.

Chen, Larry. *Sustainability Investment: The Merits of Socially Responsible Investing.* UBS Warburg, August 2001.

Cogoli, John E. *Photo-Offset Fundamentals.* Bloomington, Indiana. McKnight Puvlishing Company, 1980, 386 pp.

Comparato, Frank E. *Chronicles of Genius and Folly: R. Hoe & Company and the Printing Press as a Service to Democracy.* Culver City: Labyrinthos, 1979, 846 pp.

Conover, Theodore E. *Graphic Communications Today.* St. Paul, Minnesota: West Publishing Co., 1985, 473 pp.

Core, Erin. "Printers Weigh: Remote Proofing Options." *Graphic Arts Monthly,* May 2003, p. 32.

Cross, Lisa. "Hard Thinking about Soft Proofing." *Graphic Arts Monthly,* January 1999.

Crow, Wendell C. *Communication Graphics.* Englewood Cliffs, New Jersey: Prentice-Hall, 1986, 322 pp.

Davidson, Eric A. *You Can't Eat GNP: Economics As If Ecology Mattered.* Perseus Publishing, 2000.

Dennis, Ervin A., and John D. Jenkins. *Comprehensive Graphic Arts.* Indianapolis: Bobbs-Merrill Educational Publishers, 1983, 605 pp.

Denton, Craig. *Graphics for Visual Communication.* Dubuque, Iowa: Wm. C. Brown Publishers, 1992, 383 pp.

DeSimone, Livio, and Frank Popoff. *Eco-Efficiency: The Business Link to Sustainable Development.* Cambridge, Massachusetts: MIT Press, 1997.

Dixon, Frank. "Financial Markets and Corporate Environmental Results." Innovest working paper, 2002.

Dodd, Robin. *From Gutenberg to OpenType, An Illustrated History of Type from the Earliest Letterformns to the Latest Digital Designs.* Vancouver, Dublin, Amsterdam: Hartley & Marks Publishers, 2006, 192 pp.

Dodt, Lorette C. *Graphic Arts Production.* Homewood, Illinois: American Technical Publishers, Inc., 1990, 302 pp.

Dowell, Glen, Stuart Hart, and Bernard Yeung. "Do Corporate Environmental Standards Create or Destroy Market Value?" *Management Science,* August 2000.

Durning, Alan. *How Much Is Enough? The Consumer Society and the Future of the Earth.* New York: W. W. Norton, 1992.

Econnections: Linking the Environment and the Economy. December 1997, *Statistics Canada* No. 16-505-GPE, Ottawa, Ontario.

Ekins, Paul. *Economic Growth and Environmental Sustainability: The Prospects for Green Growth.* New York: Routledge, 2000.

Eldred, Nelson. *Package Printing.* Plainview, New York: Delmar Publishing Co., Inc., 1993, 508 pp.

Farance, Frank. "Standards Activities in Metadata," 1999–01, Farance/Edutool, www.farance.com

Farrell, Alex and Marueen Hart. "What Does Sustainability Really Mean? The Search for Useful Indicators." *Environment*, Vol. 40(9), November 1998: pp. 4–9, 26–31.

Field, Gary G. *Color and Its Reproduction*. Pittsburgh: Graphic Arts Technical Foundation, 1988, 379 pp.

Field, Gary G. *Color Scanning and Imaging Systems*. Pittsburgh: Graphic Arts Technical Foundation, 1990, 309 pp.

Field, Gary G. *Tone and Color Correction*. Pittsburgh: Graphic Arts Technical Foundation, 1991, 168 pp.

Fraser, Bruce, Chris Murphy, and Fred Bunting. *Real World Color Management*. Berkeley, California: Peachpit Press, 2005, 608 pp.

Friedman, Milton. "The Social Responsibility of Business Is to Increase Its Profits." *New York Times Magazine*, September 13, 1970.

Heal, Geoffrey. *Valuing the Future: Economic Theory and Sustainability*. Columbia University Press, 1998.

Henderson, H. *Beyond Globalization: Shaping a Sustainable Global Economy*. West Hartford, Connecticut: Kumarian Press, 1999.

Hinderliter, Hal. "The New Contract Proof." *American Printer*, February 1, 2003.

Hird, Kenneth F. *Offset Lithographic Technology*. South Holland, Illinois: The Goodheart-Willcox Co., 1995, 720 pp.

Hoffman-Falk, Marieberthe (ed.). *Digital Printing*. Océ Printing Systems, 2005, 432 pp.

Holliday, Charles O. Jr., Stephan Schmidheiny, and Philip Watts. *Walking The Talk, The Business Case for Sustainable Development*. Greenleaf Publishing, September 2002.

Howe, Walt. *A Brief History of the Internet*. August 2016, www.walthowe.com/navnet/history.html.

ISEA. AccountAbility 1000—A Foundation Standard in Social and Ethical Accounting, Auditing, and Reporting. London: Institute of Social and Ethical Accountability, 1999.

Kleper, Michael L. *The Handbook of Digital Publishing, Volumes I and II*. Upper Saddle River, New Jersey: Prentice Hall, 2001, 1,004 pp.

Larish, John. *Digital Photography*. Torrance, California: Micro Publishing Press, 1992, 208 pp.

Leiner, Barry M., Vinton G. Cerf, David D. Clark, Robert E. Kahn, Leonard Kleinrock, Daniel C. Lynch, Jon Postel, Larry G. Roberts, and Stephen Wolff. *A Brief History of the Internet*. Reston, Virginia: Internet Society, 2005.

Leonard-Barton, Dorothy, and William A. Kraus. "Implementing New Technology." *Harvard Business Review*, November–December 1985, pp. 102–110.

Levarie, Norma. *The Art & History of Books*. New York: Da Capo Press, 1968, 315 pp.

Levenson, Harvey R. "Electronic Digital Photography." *1994 Technology Forecast*. Pittsburgh: Graphic Arts Technical Foundation, 1994, pp. 10–12.

Levenson, Harvey R. "From McLuhan to Wilkens: Bridging the Technologies of Design, Print, and Telecommunications at Cal Poly." *The Prepress Bulletin*, July/August 1985.

Levenson, Harvey R. *Complete Dictionary of Graphic Arts and Desktop Publishing Terminology*. Thousand Oaks, California: Summa Books, 1995, 271 pp.

Levenson, Harvey Robert. *Understanding Graphic Communication*. Pittsburgh: GATFPress, 2000, 248 pp.

Levy, Uri, and Gilles Biscos. *Nonimpact Electronic Printing*. Charlottesville, Virginia: InterQuest, Ltd., 1993, 314 pp.

Lichty, Tom. *Design Principles for Desktop Publishers*. Belmont, California: Wadsworth Publishing Company, 1994, 240 pp.

Lipetri, Joe. "Remote Proofing Delivers." *American Printer*, September 1, 2001.

McDonough, William, and Michael Braungart. *Cradle to Cradle: Remaking the Way We Make Things*. New York: North Point Press, 2002.

Meadows, Donella H., Dennis L. Meadows, and Jorgen Randers. *Beyond the Limits*. Toronto: McClelland & Stewart Inc., 1992.

Meadows, Donella H. et al. *The Limits to Growth*. Washington D.C.: Potomac, 1972.

Miley, Michael. "The Ties That Bind." *Electronic Publishing*, October 2004.

Molla, R. K. *Electronic Color Separation*. Montgomery, West Virginia: R. K. Printing & Publishing Company, 1988, 288 pp.

Mort, Richard A. *How to Save a Bundle on Printing*. Portland, Oregon: Richard A. Mort, 1989, 149 pp.

National Association for Printing Leadership, www.napl.org

Negroponte, Nicholas. *Being Digital*. New York: Random House, 1995, 255 pp.

Nelson, Roy Paul. *The Design of Advertising*. Dubuque, Iowa: Wm. C. Brown Company Publishers, 1962, 303 pp.

Nothmann, Gerhard A. *Nonimpact Printing*. Pittsburgh: Graphic Arts Technical Foundation, 1989, 110 pp.

Oakley, A. L., and A. C. Norris. "Page Description Languages: Development, Implementation, and Standardization." *Electronic Publishing*, September 1988, Vol. 1(2), 79–96.

Parker, Roger C. *Looking Good in Print*. Scottsdale, Arizona: Paraglyph Press, 2006, 334 pp.

Parsons, John. "Easy Dot-Com, Easy Dot-Go." *The Seybold Report*. Media, Pennsylvania: Seybold Publications, May 7, 2001

Parsons, John. "Printing in the Green." *The Seybold Report*. Media, Pennsylvania: Seybold Publications, February 4, 2002.

Parsons, John. "The X-Files." *The Seybold Report*. Media, Pennsylvania: Seybold Publications, June 18, 2001.

Pfiffner, Pamela, Bruce Fraser, and Dave Feasey. *How Desktop Publishing Works*. New York: Ziff Davis Publishers, 1994, 198 pp.

Pfiffner, Pamela. *Inside the Publishing Revolution: The Adobe Story*. San Jose, California: Adobe Press, 2002, 256 pp.

Polishuk, Tom (ed.). "Combine and Conquer." *Package Printing*, October 8, 2004, 3 pp.

Ponting, Clive. *A Green History of the World*. Penguin, 1991.

Printing Industries of America. *The Value of Print*. Warrendale, Pennsylvania, 2012, 70 pp.

Prust, Zeke A. *Graphic Communications: The Printed Image*. South Holland, Illinois: The Goodheart-Willcox Co., 1989, 544 pp.

Rea, Douglas Ford. "Electronic Still Photography," ESP '93 Teleconference Program Notes. Rochester, New York: Rochester Institute of Technology, April–May 1993, pp. 30–37.

Reid, David. *Sustainable Development: An Introductory Guide*. London, UK: Earthscan Publications, Ltd., 1995.

Rigsby, Lana and Eagleman, David. *A Communicator's Guide to The Neuroscience of Touch*. Boston: Sappi North America, 2017, 54 pp.

Romano, Frank J. *Digital Printing*. San Diego: Windsor Professional Information, LLC, 2000, 262 pp.

Romano, Frank J. *History of the Linotype Company*. Rochester, New York: RIT Press, 2014, 480 pp.

Romano, Frank J. *History of the Phototypesetting Era*. San Luis Obispo, California: California Polytechnic State University, 2014, 340 pp.

Romano, Frank J. *Machine Writing and Typesetting: The Story of Sholes and Mergenthaler and the Invention of the Typewriter and the Linotype*. Salem, New Hampshire: GAMA Communications, 1986, 146 pp.

Romano, Frank J., and Richard M. Romano. *GATF Encyclo pedia of Graphic Communications*. Pittsburgh: GATFPress, 1998, 945 pp.

Rosner, Charles. *Printer's Progress, A Comparative Survey of the Craft of Printing, 1851-1951*. Cambridge: Harvard University Press, 1951, 125 pp.

Sappi North America. *The Standard*. Vol. 1-6. https://www.sappi.com/the-standard

Segal, Ben, CERN IT-PDP-TE. "A Short History of Internet Protocols at CERN," April 1995, http://ben.home.cern.ch/ben/TCPHIST.html

Sharma, Abhay. *Understanding Color Management*. Boston: Cengage Learning, 2003, 384 pp.

Solow, Robert M. (1991). *Sustainability: An Economist's Perspective*. Reprinted in Stavins, Robert N. (2000). *Economics of the Environment: Selected Readings, fourth edition*. W.W. Norton & Company: 131–138.

Spilker, J.J. *Digital Communications by Satellite*. New Jersey: Prentice Hall Inc., 1982.

Todoroki, Emiko. *Globalization and Corporate Social Responsibility*. The World Bank Institute, May 2002.

Walker, John R. *Graphic Arts Fundamentals*. South Holland, Illinois: The Goodheart-Willcox Co., 1980.

Webb, Joseph W. and Richard M. Romano. *The Third Wave*. Lowell, Massachusetts: King Printing, 2014, 174 pp.

Wentzel, Fred, Ray Blair, and Tom Destree. *Graphic Arts Photography: Color*. Pittsburgh: Graphic Arts Technical Foundation, 1987, 151 pp.

White, Alan T. "Putting a Price on Nature." Earthwatch Institute Journal, August 2000. Reprinted in *Ocean Seas: The Online Magazine for Sustainable Seas*, October 2000, Vol. 3(10).

World Business Council for Sustainable Development. *Eco-efficiency and Cleaner Production: Charting the Course to Sustainability*. Geneva: WBCSD, 1994. 17 pp.

World Commission on Environment and Development. *Our Common Future*. Oxford, UK: Oxford University Press, 1987. Xerox, http://www.xerox.com/PARC/dlbx/library.html

Zipper, Bernd. *pdf + print 2.0 – PDF Handbook for prepress industry*. Media, Pennsylvania: Seybold Publications, 2002, 242 pp.

Zipper, Bernd and John Parsons. "PDF Collaboration in Action." *The Seybold Report*. Media, Pennsylvania: Seybold Publications, September 3, 2001.

Index

Symbols

3D
  die-cutting 171
  packaging design 111
  visualization software 180
42-line Bible 35

A

Adhesive 166
Adobe
  Bridge 226
  Creative Cloud 86, 221
  Creative Suite 104
  Flash 222, 223, 224
  Illustrator 86, 87, 106
  InCopy 224
  InDesign 86, 88, 90, 93, 94, 221, 223, 224, 227, 230
  InDesign Server 86, 223, 224
  OPEN 96
  Photoshop 87, 104, 227
  PostScript 85, 88, 91, 151, 221
Advanced Research Projects Agency Network 217
Advertising 56, 204
Africa 177
Agfa 109, 153
Aldus PageMaker 82, 85, 86, 221
Allen Datagraph 181
Al-Radaideh, Bassam 20
Amazon Kindle 219
America 11, 28, 40, 41, 53, 84, 120, 122, 146, 152, 156, 177, 179, 187, 189, 195
American National Standards Institute (ANSI) 120, 201
American Telephone and Telegraph Company 213
Andreessen, Marc 218
Anilox roll 142, 143
Apple Computer 47, 85
Asia 32, 34, 38, 45, 177, 189
Assembly process 165, 169
Association for Print Technologies 201
Automation 75, 80, 95, 96, 160

B

B65 Committee 202
Bank stationery products 69
Barbaran 183, 188
Barnes & Noble 71, 73
Basic bindery 169
Baskerville, John 38, 39
Bavarian limestone 45
Berners-Lee, Timothy 212, 217, 218
Bestcolor 109
Bindery 89, 160, 171, 202
Binding 160, 166, 195
Blanket 136, 140, 147, 155, 156
Bleed 89, 90, 95
Bobst 183, 188
Bodoni, Giambattista 38, 39
Bone folder 164
Book 10, 11, 13, 41, 71, 168, 171, 204, 228
Book Industry Study Group 228
Boxing 169
Brand 171
Brightness 122
Broadband 54
Buckle folder 164, 165
Bullock, William 43

Bush, Vannevar 217

Business 56, 69, 70, 193, 195, 225

Business forms and bank stationery printing 69

Business-to-Business (B2B) 225

Business-to-Customer (B2C) 225

C

Calendering 118, 119

Caliper 120, 122

Campbell, John 41

Canon/Océ 153, 181

Carli, Don 124

Carlson, Chesley 46, 47

Cartons 181

Cascading Style Sheets 218

Case-bound cover 166, 167, 168, 169

Caslon, William 38, 39

Caxton, William 39

Cellulose 117, 125

CERN 217

CGATS 202, 204

CGS-ORIS 109

Checks and deposit/withdrawal forms 69

CIEL*a*b* 104

CIP4 96, 202

Clickable Paper 11, 12, 13, 219

Cloud 86, 221

CMYK 22, 87, 89, 94, 95, 101, 102, 105, 106, 107, 108, 127, 134, 154

Coatings 121

Collating 165

Color 70, 71, 83, 84, 87, 97, 98, 99, 100, 102, 103, 104, 105, 106, 108, 122, 127, 154, 199, 203, 204

Color Exchange Format 106, 127

Combination printing 146

Commercial printing 66, 67, 68

Computer and specialty forms 69

Computer graphics 52

Computer Integrated Manufacturing 205

Computerized page layout 221

Computer Numeric Controlled 171, 205

Computer-to-plate 152

Conductive metals 213

Consumer products 177

Consumer WtP storefronts 225

Content Management Systems 230

Converters 180

Copperplate engravings 37

Copyrights 200

Corel Draw 86

Corrugated board printing 183

Covers 160, 168

Cover stock 168

Creative Edge Software 111

Currency 74

Customers 54, 55

Customer Service Representative 194

Cutting 160, 161, 162, 172, 202

D

Dampening fountain 139

Dandy roller 118

Database publishing 219

Day, Matthew 41

Daye, Stephen 40

Debossing 170, 171

Defense Advanced Research Projects Agency 216

De-inking 120

Delivery 199

Demographic
   binding 169, 170
   groups 15, 72, 88, 179
   printing and distribution 73

Densitometer 136

Density 136

Design Reproduction Technologist 22, 23

Desktop
  Design 85, 92
  Publishing (DTP) 48, 82

Diamond Sutra 34

Die-cutting 171, 204

Digital 15, 18, 21, 48, 54, 56, 75, 76, 80, 83, 87, 90, 93, 114, 126, 128, 129, 132, 133, 147, 148, 149, 151, 152, 153, 154, 157, 160, 161, 171, 172, 176, 177, 179, 180, 181, 182, 184, 188, 209, 210, 214, 226, 228, 229, 230
  Digital asset management 224, 225
  Digital media 15, 48, 229
  Digital package printing 76
  Digital press 182
  Digital printing 54, 56, 133, 147, 148, 153, 154, 180, 184
  Digital workflow 153

Digital Subscriber Line 214

Direct-to-plate 179

Disclaimer of Express Warrantee 200

Domain Name System 217

Donald Knuth, Donald 221

Donkin, Bryan 33

Dot gain 136

Dow Jones 25, 52, 215

DRUPA 157, 179

Dupont 108, 123

Durst 183

E

eBook 11, 13, 219

EFI 109, 153, 188

E Ink Corporation 217

Electronic 21, 28, 36, 46, 174, 198, 218
  Electronic binderies 174
  Electronic color separation scanners 36

Electrophotographic 47, 129, 150, 151

Electrostatic 150

Email 216

Emboss 162, 168, 169, 170, 171

Engelbart, Douglas 217

Engraving 140

Enterprise Resource Procurement 225

Environmental refulations 118

Environmental Protection Agency 58

ePac 187

Erasmus 40

Esko Graphics 96, 111

Esquire magazine 219

Estimate 195

Europe 33, 35, 39, 41, 141, 177, 189

F

FDA regulations 182

Federal Communication Commission 214

FedEx 67

Felt 118

Fiber 120, 125

File Transfer Protocol 218

Film 84, 88

Financial and legal printing 74

Finishing 159, 160, 169, 170, 171, 189

Flatbed 138

Flexographic printing 143

Flexography 142, 144

FOGRA Graphic Technology Research Association 203

Foil 170

Foil stamping 170

Folding 160, 163, 164, 181

Folding carton 181

Folding dummy 90

Forest Stewardship Council 124, 125

Fotosetter 44

Fountain solution 136, 137, 138, 139, 140, 143
Fourdrinier, Seasly 33, 117, 118
Franklin, Benjamin 40, 41
Freeman's Oath 40
Fujifilm 108, 146, 153, 182

G

G7 98, 101, 107
Gannett 36, 72
Gathering 165
GCR 106
General Electric Corporation 56
GE/TAPPI specification 122
Ghent Workgroup 88
Gloss 121
Gordon, George Phineas 43
Goudy, Frederick W. 38, 39
GRACol 107
Graphic Communications Association 203
Gravure 44, 140, 141, 142
Grayscale 107
Greeting card 70
GretagMacbeth 106
GTO-DI 153
Guillotine cutter 162, 163
Gutenberg, Johann 15, 20, 28, 29, 30, 32, 33, 34, 35, 36, 37, 38, 39, 40, 41, 45, 46, 48, 52, 72, 81, 134, 138, 146, 157, 161, 211, 212, 220

H

Hachette 71
Halftone 83, 90, 92
Haptic 13, 73, 122, 226, 227, 228
Harper's 42
Harris, Benjamin 41, 43, 44, 52
Harris Corporation 52
Harvard Business School 56

Hearst 219
Heatset inks 140
Heidelberg 134, 146, 153
Hewlett Packard 153
Hollander beater 33
Homer 40
Human perception 103
Hypertext Markup Language 218
Hypertext Transfer Protocol 217

I

ICC profile 104, 105, 107, 203
Idealliance 107, 203, 205
Illustration 80, 86, 107
Image assembly 84
Imposition 84, 90
Inca 183
Indemnification 200
InDesign Markup Language 223
Ingram 71
Ink 37, 89, 113, 114, 123, 126, 127, 138, 139, 202, 219
Inkjet 98, 109, 110, 128, 147, 149, 150, 172, 180
  Continuous drop 150
  Drop-on-demand 151
  Inkjet printing 149, 150
  Inkjet proofing 109, 110
Inline finishing 170
In-plant printing 76
Inserting 165, 169
Intaglio 141
Integrated Services Digital Network 214
International Color Consortium (ICC) 104, 203
International Commission on Illumination (CIE) 104

International Standards Organization (ISO) 119, 203

Internet 25, 36, 48, 52, 54, 70, 86, 154, 177, 210, 211, 212, 213, 214, 215, 216, 217, 218, 219, 220, 221, 222, 224, 226, 228, 229

Internet Protocol 217

Invoicing 197

IT8 204

J

Janson, Nicholas 38, 39

Japan 34, 178

Joachim, Carl 187

Job Definition Format 96, 162, 172, 205

Job Messaging Format 205

K

KBA 153

Kleinrock, Leonard 217

Knife folders 164

Knockout 94

Knott, Jack 187

Kodak 108, 146, 153, 154, 182
　　Kodak Approval 108

Koenig, Friederich 41, 142

Komori 153

Konica Minolta 153, 188

Korea 34, 214

L

Label printing 181, 182

L*a*b* values 106

Lamination 189

Landa 153, 156, 181, 182, 188

Leonard-Barton, Dorothy 56

Letterpress 137, 138

Liability 200

Licklider, Joseph 216

Lighting conditions 103

Lightroom 227

Linotype 30, 35, 36, 43, 52, 54, 83

Lithography 28, 44, 45, 139, 203

Local Area Networks 217, 220

London Times 41, 42

M

Machines Dubuit 184

Macintosh 47, 85

Magazine 73

Mahon, Charles 43

Makeready 155, 156, 157, 199

Management Information System 205

manroland 134, 153

Manufacturing 114, 116, 117, 126, 205

Manutius, Aldus 39, 40

Marconi, Guglielmo 214

Mark Andy 143, 188

Marketing 57

Matte 121

Mechanical 167, 168

Megenthaler, Ottmar 43

Memjet 153, 183

Metadata 224, 225

Metal 37, 184

Metallic foil stamping 169, 171

Mexico 40

Middle East 177

MIT 33, 216, 217, 218

MIT Media Lab 33

Mitsushita Corporation 52

Mobile 219

Moiré 91, 92

Monotype 43

Moore, Geoffrey 51

Morse, Samuel 213

Mosaic 218

Movable type 34, 35

N

Nanography 156

National Association for Printing Leadership 195

National Center for Supercomputing Applications 218

Navigator 218

Near-field Communication 186

Negative film 77, 83, 144, 198

Nelson, Ted 217

Net Neutrality 214

Netscape 218

Network evolution xi

New England 40, 41

New Testament 40

New York Tribune 43

Non-heatset 140

North American Industry Classification System 66

O

Occupational Safety and Health Act 58

Offset lithography 126, 135

Omet 188

On-demand printing 67, 74

Online ordering 224

Online technology 212

On-screen renderings 222

Open Prepress Interface 94

Optical 121, 214
    Optical brighteners 121

Optical fiber cables 214

Order 196

Original Equipment Manufacturer (OEM) 128, 129, 157, 172, 173, 180, 182, 183, 185, 187, 188, 189

Overprint 94

Oxford 217

P

Pablos, Juan 40

Package printing 76, 146

Packaging 64, 76, 87, 111, 171, 175, 176, 177, 178, 179

Page layout 83, 86, 95, 230

Pantone 105, 106

Paper 10, 11, 12, 13, 48, 71, 113, 114, 115, 116, 117, 118, 119, 120, 121, 124, 164, 172, 188, 218, 219

Papermaking 118

Papyrus 32

PDF 25, 80, 86, 88, 89, 90, 91, 93, 94, 95, 106, 108, 109, 151, 221, 222, 223

PDF/X 80, 86, 88, 89, 90, 93, 94, 95, 106, 109

Penguin 71

Pennsylvania Gazette 41

Personalization 185, 186

pH 136

Phaestos disc 32

Pigment 70, 103, 105, 121, 126, 156

Pi Sheng 34

Pitney Bowes 153

Pixel 87

Plastic 167

Platemaking 84

Poor Richard's Almanac 40

Postpress 159, 160, 161, 162, 170

PostScript 85, 88, 91, 151, 221

Preflight 93

Prepress 76, 79, 80, 94, 199, 202, 204

Primera 181
Printing Industries of America 11, 53, 179, 195
PrintTalk 203
Process color 155
Production 124, 179, 196, 204
Promotional item printing 69
Proofing 97, 98, 108, 110, 143, 199
Proposal 195
Protestant Reformation 35
Publication printing 70
Publisher 86
Punched paper tape 43

Q

Quality and productivity 201
QuarkXPress 86, 88, 90, 93, 94, 221, 224, 230
Quick Response (QR) Code 11, 12, 14, 186
Quotation 195

R

Radio 169, 186
Radio Frequency Identification 169, 186
Ramus, Peter 36
Random House 71
Raster Image Processor 91
Ready, Set, Go! 86
Registration 135
Renaissance 22, 28, 39, 40
R. Hoe & Co. 41, 42
Ricoh 11, 12, 153, 154, 181, 219
Right-reading 137, 138, 141
RISO 153
Rittenhouse, William 41
Robert, Louis 33
Roberts, Lawrence 216, 217
Roland 110
Roman type 39
Rosette pattern 91

Rotary letterpress 138, 139
Rotary sheetfed gravure 141
Rotogravure 142
RR Donnelley 96
Rubel, Ira 45
Rust, Samuel 41

S

Saddlebar 165
Saddle stitching 166, 167, 168, 174
Safety 58, 118, 202
Satellite 214
Schoeffer, Peter 35
School yearbook xiv
Science of touch, see Haptic
Scitex 52, 109, 136, 153, 183, 188
Scoring 162, 189
Scratch-off gaming 135
Screen Americas 153
Screen-printing 144
Securities and Exchange Commission xiv
Security printing 74
Self-adhesive coating 182
Self-cover 16, 168
Senefelder, Alois 44, 45, 48, 211
Server 86, 223, 224
Seybold, John W. 221
Sheetfed 71, 107, 122, 134, 140, 141, 155, 173, 174, 203, 204, 205
Shilling, Pavel 213
Shotoku 34
Side guides 162
Side-sewn binding 167
Siemens 213
Signature 163, 165, 166, 168, 169, 173

Sign printing 68
Simon and Schuster 71
Smart packaging 184, 186
Smithers Pira 177, 179, 180, 183
Smith, Peter 41
Smythe sewing 167
SNAP 204
Soft proofing 111
Software-as-a Service 221
Solvents 143
South Korea 214
Special effects 171
Specialized presses 142
Specialty Graphic Imaging Association 68
Specialty printing 68
Specialty publication xv
Specification for Web Offset Publications (SWOP) 205
Specifications for Newsprint Advertising Production (SNAP) 204
Spectrophotometer 104, 107, 111, 136
Spot color 95, 105, 106
Spot varnish 170, 189
Stanford 217
Stochastic screening 92
Stone 117
Storage 200
Substrate 28, 32, 103
Subtractive color 103
Sustainability 114, 124
Sustainable Forestry Initiative (SFI) 125
Syncom 215
Synthetic 124

T

Tack 127
Ta-jong 34

Taxes 200
Teamwork 193
Technological change 54
Telecommunication 210
Telegraphy 213
Television 12
Template 222
Template-based workflows 222
TEX 221
Text transmission 213
Three-knife trimmer 163, 173
Till 188
Tints 90, 92
Toner 113, 114, 126, 129, 150
Total Ink Coverage 89
Transmission Control Protocol 217
Trap 94, 135, 136
Trim 89
Ts'ai Lun 32, 33, 36, 48, 211
T-shirts 145
Type 32, 38, 43, 83, 86
Typesetting 83, 93
Typography 38
Tyvek 123

U

UCLA 217
UC Santa Barbara 217
Ultraviolet 152
United States Government Printing Office 76
University of Utah 217
USA Today 215
UV-curable ink 145
UV curing 145

V

Variable data printing (VDP) 72, 75, 146, 148, 149, 170, 172, 179, 184m 185, 220

Ventura Publisher 86

Versions 222

ViaStone 123

Video 14

W

Wall Street Journal 25, 52, 215

Wang Chen 34

Washington Hand Press 42

Water 125, 152

Water-based latex inks 152

Web authoring 22

Web-based document composition 222

Web browser 218

Web communication 54

Web-enabled printing 54

Websites 15, 227

Web storefronts 225

Web-to-Print 61, 220, 222, 224, 225

What You See Is What You Get (WYSWYG) 47, 85, 224

Wide-format digital printers 151

Wilkins, Jim 53

World Wide Web 154, 211, 216, 217, 218

World Wide Web Consortium (W3C) 218

Wrong color mode 93

X

Xanté 153, 183

Xeikon 153, 155, 181

Xerography 129, 151

Xerox 33, 85, 152, 153, 154, 218, 221

Xerox Palo Alto Research Center (Xerox PARC) 33, 216, 219

XML 205, 230

X-Rite 106

Xyvision 221

Y

Yupo Corporation 123